P9-DHI-293

Newton's Clock
Chaos in the Solar System

Newton's Clock

Chaos in the Solar System

•

IVARS PETERSON

W. H. Freeman and Company
NEW YORK

Library of Congress Cataloging-in-Publication Data

Peterson, Ivars.
Newton's clock : chaos in the solar system / Ivars Peterson.
p. cm.
Includes bibliographical references and index.
ISBN 0–7167–2396–4 :
1. Celestial mechanics. 2. Chaotic behavior in systems.
3. Ephemerides. I. Title.
QB351.P48 1993
521—dc20 93-7176
CIP

PRINTED IN THE UNITED STATES OF AMERICA

1 2 3 4 5 6 7 8 9 0 VB 9 9 8 7 6 5 4 3

In memory of my father,

ARNIS PETERSON

Contents

•

Preface

•

The moon was shining sulkily,
Because she thought the sun
Had got no business to be there
After the day was done—

LEWIS CARROLL, *Through the Looking-Glass*

THE NAMES of the places of my childhood still echo through my mind: McKenzie Island, Red Lake, Minaki, Caribou Falls, Kenora—small communities overshadowed by the immense, enshrouding wilderness of northwestern Ontario. Etched equally sharply into my memory is the velvety black cloak of a night sky undisturbed by human light. Strewn with mysteriously glistening specks, this vast, delicately decorated canopy inspired both awe and wonder.

I can't remember my first impression as a young child of the night sky. All too quickly, I learned about stars, and planets, and moons, and it became a matter of faith to associate those points of light in the sky with distant, massive, rapidly moving objects. One mystery replaced by another.

Arthur Loeb, a student of the visual arts and a professor at Harvard University, can describe his reaction to his first encounter with the night sky because his parents told him what he said. "It's full of holes," the child remarked.

There's no single obvious way to interpret the evident contrast between light and dark we see in the sky. Children learn from their parents and teachers what to see. Step by step, we assemble the observational and mathematical scaffolding that implies a whirling Earth, a revolving solar system, a drifting galaxy, an expanding universe.

Given the huge reservoir of knowledge on which we now build our vision of the cosmos, it's perhaps difficult to imagine how, over several millennia, ancient observers and their successors painstakingly transformed observations of spots of light, or even holes in the firmament, into a theory of suns and planets and moons. The clues they pondered came from the curiously varied movements of these bright specks—movements that more often than not displayed subtle variations apparent only to patient, inspired observation.

In an age when the night sky pales in the glare of artificial light and the distractions of human activity, many of us have lost our sense of the rhythms evident in the sky. But it's possible to recover a semblance of this celestial clockwork and to appreciate the rich store of information available to persevering observers, past and present.

When my son Eric was two years old, we used to go for walks in the early evening before bedtime. Wandering along the lane behind our house, we would look at the flowers peeking through the fences lining the pavement, sample the pebbles on the roadbed, collect pine cones, and stop to yell a greeting into the dark cavity below the drainage grating. But it was the moon at dusk that most attracted Eric's attention. From the first time that he became aware of its presence in the sky, he would always seek out this strangely inviting apparition. He took particular delight and pride in being the first to catch sight of the moon on any given evening.

But the moon wasn't always there, and its appearance changed from week to week. Sometimes it would be visible well before dusk, and at other times after dawn. Sometimes it was low on the horizon, at other times high in the sky. To anticipate Eric's questions, I began to think about how the moon moves across the sky and to construct my own mental model of its motion so that I could predict where it was likely to be. Of course, I could have looked all this up in any one of innumerable astronomy books or almanacs, or even checked a calendar that marks the lunar phases, but I didn't want to take such an easy way out. I joined the spirit of the ancient observers, patiently tracking the moon's excursions and looking for patterns on which I could hang my predictions. I was struck by how much I could learn and deduce from even an informal perusal of movements across the sky. The remarkable achievements of the ancient astronomers, using nothing more than the naked eye, no longer seemed so foreign and magical.

In this book, I have attempted to tell the extraordinary story of the human search for an understanding of the motions of the moon and the planets against their starry backdrop. It makes for an amazing mathematical detective story. For it was in the interplay among astronomy, physics, and mathematics that our vision of the solar system evolved, from descriptive theories to laws of motion and finally to the intrinsic uncertainties embedded in mathematics itself. What we see in this panorama across the ages is a relentless search for order and harmony. But the dream of celestial certainty has been derailed by the limitations of the mathematics. What we end up with is the startling emergence of chaos in the celestial clockwork.

Indeed, we face a growing realization that there is a real limit to what we can know and understand about a solar system that happens to have just the right characteristics to have kept us searching for so long. A solar system simple enough to allow—and even encourage—us to search for patterns and deduce grand, universal laws; a solar system just complicated enough to keep us wondering without stopping us in our tracks. In many ways, our studies of the solar system represent a spectacular success story predicated on the peculiar characteristics of the platform from which we view the cosmos.

We know Earth's shuddering spin so well that we can nonchalantly add a leap-second to our calendars whenever necessary, and we can bring a spacecraft to a pinpoint rendezvous years and billions of kilometers from Earth. Yet the solar system's future billions of years from now and the future of any particular planet remain hidden in our clouded mathematical crystal ball. Physics and mathematics somehow conspire to keep us from discovering the ultimate answers.

I have not tried to write a definitive, comprehensive history of celestial mechanics. Rather, by highlighting important individuals in the field and their achievements, I've attempted to illustrate our progress in celestial mechanics and to place in historical context contemporary research on dynamical systems and chaos. I have been particularly fortunate to have had the benefit of a considerable and steadily growing body of scholarly work in the history of science. Several studies reveal new insights about such figures as Johannes Kepler, Isaac Newton, and Henri Poincaré, which sometimes quite strikingly contradict the information commonly presented in textbooks and popular writings.

The historical studies also help remind us of the rich intellectual environments in which the major figures in celestial mechanics have worked. Newton and his contemporaries, for example, could perform remarkable feats of mathematical prestidigitation to solve a variety of problems that would stymie most modern mathematicians. Only when we try to read their works by placing ourselves in their era can we really appreciate the power and subtlety of the arguments they brought to bear to persuade fellow scholars of the value of their radical theories.

We often fail to credit astronomy as a highly mathematical pursuit. The need to predict and pinpoint astronomical phenomena played a major role in driving the development of mathematics, just as mathematical discoveries vastly expanded the domain of astronomy. Few professionals greeted the invention of time-saving logarithms as warmly as the astronomers, who were thereby able to save many precious days in their painfully lengthy, complex calculations of planetary and lunar positions. Indeed, before its modern incarnation, the word *computer* signified not a calculating machine but a human being assigned the job of calculating, and no astronomer could do his job without a bevy of clerk "computers" to perform the necessary calculations. Today, both amateur and professional astronomers rely heavily on sophisticated electronic computers to do similar tasks.

Because I am neither a historian nor an expert in celestial mechanics, astronomy, and mathematics, I have relied on the work of a host of historians, astronomers, and mathematicians to guide my writing. I've gleaned ideas, examples, and insights from numerous lectures, interviews, published papers, articles, and conversations. References that I found particularly rewarding and useful are included among the books and articles listed in the bibliography at the end of the book.

I am grateful to Owen Gingerich, Philip Holmes, and Gail S. Young, who were kind enough to read the manuscript, point out any problems, and suggest improvements. Any errors that remain, however, are my responsibility.

I also wish to thank the following individuals for helping me by explaining ideas, providing reference material, suggesting sources of information, and pointing out details I had initially overlooked: Ken Brecher, Robert Brumbaugh, Joseph Burns, David Chudnovsky, André Deprit, Martin Duncan, Sylvio Ferraz-Mello, Peter Goldreich, Louise Golland, Daniel Goroff, Keith Greiner, Martin Gutzwiller, Liam Healy, James Klavetter, Jacques Laskar, Myron Lecar, Richard McGehee, Gerald Quinlan, Bruce Stephenson, Gerald Sussman, Scott Tremaine, George Wetherill, and Jack Wisdom. My apologies to anyone I have inadvertently omitted from this list.

For help in locating illustrations for use in this book, I greatly appreciate the efforts of Ruth Freitag of the Library of Congress; Deborah Warner, Steven Turner, Uta Merzbach, and Peggy Kidwell at the Smithsonian Institution's National Museum of American History; Karin Lindberg of the Mittag-Leffler Institute; Richard Dreiser of the Yerkes Observatory; Chandra Wilds of the MIT News Office; Kate Desulis of the Adler Planetarium; and Scott Fields of the University of Wisconsin-Madison.

It is no mean feat to turn a thick manuscript and an array of illustrations into a finished book. My thanks to publisher Jerry Lyons, project editor Diana Siemens, and everyone else at W. H. Freeman and Company for their care in turning this material into an admirable volume.

IVARS PETERSON
June 1993

Newton's Clock
Chaos in the Solar System

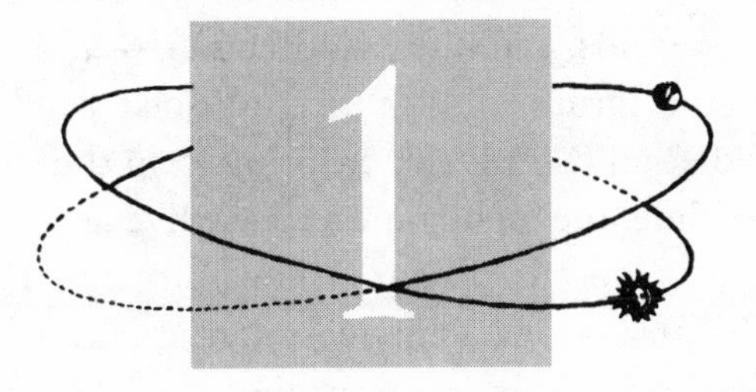

Chaos in the Clockwork

•

The heavens themselves,
the planets, and this centre,
Observe degree, priority, and place,
Insisture, course, proportion, season, form,
Office, and custom, in all line of order.

WILLIAM SHAKESPEARE
(1564–1616), *Troilus and Cressida*

SWEEPING AROUND the sun along a grand loop, the flattened ball we know as Earth hurtles through space at a breath-defying pace. Held captive by the sun's gravity, it maintains an elliptical course, every second adding another 30 kilometers to the log of its perpetual voyage. Its traveling companions maintain a respectful distance, each one held to its own well-worn track around the sun.

These nine planets respond not only to the enduring attraction of the sun but also to that of their neighbors. The competing tugs of these celestial compatriots append a restless spectrum of minute wiggles to a basic, sun-dominated motion, causing a jangle of deviations from perfect geometry. In this intricate, discordant symphony of the planets, the giants Jupiter and Saturn call out most loudly. Mercury, Venus,

Mars, Uranus, Neptune, and Pluto contribute quieter voices. But even the imperceptible, evanescent whispers of the lesser objects in the solar system—asteroids, satellites, and comets—add to the celestial chorus, as does the thrum of the fluttering solar wind of accelerated particles and radiation continually erupting from the sun.

At the same time, the globe on which we live and wonder spins on its own axis even as it orbits the sun. Like a gargantuan twirling top, it wobbles and tilts. It shudders with every earthquake and twists fitfully with every giant swirl within its atmosphere or seas. Any unevenness in its shape or in the distribution of materials making up its crust unbalances its movements and provides a lever by which the sun and other bodies can further wrench Earth from a pure and simple motion.

This modern, remarkably detailed picture of solar system dynamics represents an astonishing triumph of human reasoning. Tugged one way, then another while whirling through space, Earth serves at best as a rickety platform from which to observe and contemplate the

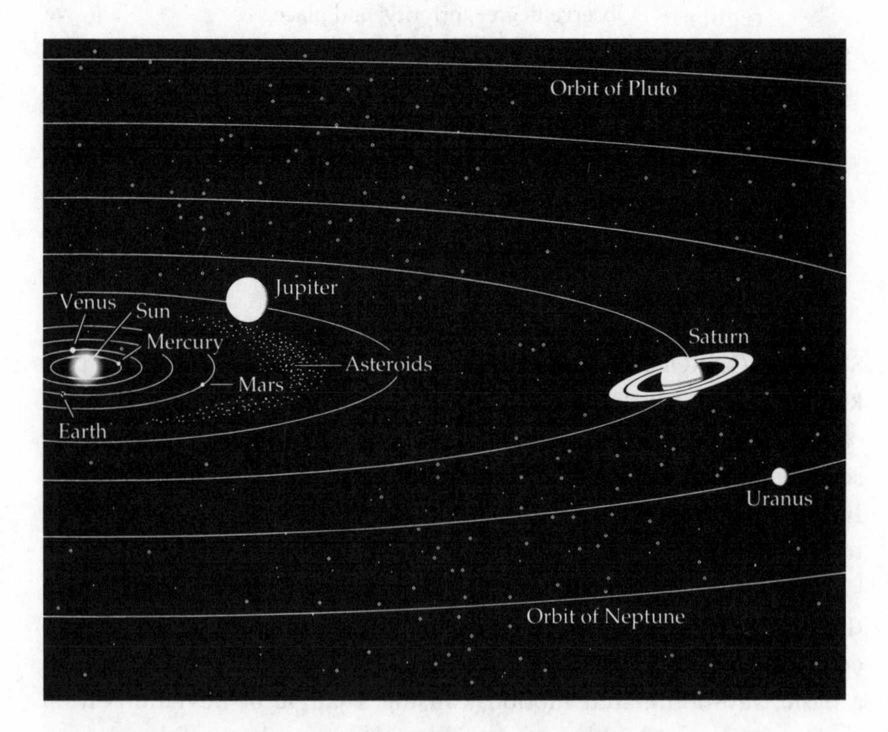

From the sun to Pluto's orbit, our solar system stretches approximately 6 billion kilometers across space.

heavens. Moreover, nothing in everyday experience tells us directly that Earth moves. What we know about the workings of the solar system we learned first from the movements across the night sky of nothing more than pinpricks of light.

Discovery starts with personal experience. It doesn't take long for a young child to accumulate sufficient empirical evidence to sense that we are somehow tied to the ground on which we first lie, then crawl, walk, and run. Objects fall. Rocks tumble down slopes. Water flows along channels to lower basins. An arrow pierces the air in a great arc, first rising but inevitably falling to Earth. These movements have a consistent, natural direction.

The unfailing, daily passages of the sun across the sky add another crucial element, as do the corresponding movements of stars at night. These cyclic, predictable motions powerfully suggest the existence of some sort of Earth-centered order that contrasts sharply with an unpredictable existence fraught with the vagaries of weather and the perplexities of human behavior. In an unruly environment, such evidence of regularity provides reassurance in the face of great unknowns.

It's hardly surprising that in days long gone, priests, scholars, poets, and astronomers focused on celestial cycles, on sequences of events in the skies that repeated themselves at soothingly regular intervals. Only in the stately progress of the stars, repeating their motions night after night with pristine precision, in the unflagging regularity of the sun's daytime trek across the sky, and in the periodic changes in the moon's appearance and position did our forebears find the clearest evidence of some sort of rational design for the bewildering universe in which they found themselves.

But closer observations over months and years reveal subtle shifts in these patterns and a variety of unexpected movements. The sun, for instance, doesn't rise at precisely the same point on the horizon every day. Instead, the location of sunrise drifts back and forth along the horizon. These recurring excursions define a longer cycle tied to the changing seasons. Similarly, particular stars rise at different locations along the horizon and, at certain times of the year, disappear entirely from the sky for lengthy periods. These movements also have a definite rhythm tuned to the seasons.

The moon has its own cycles and thus provides another natural unit of time. Like the sun, it rises and sets at different points along the

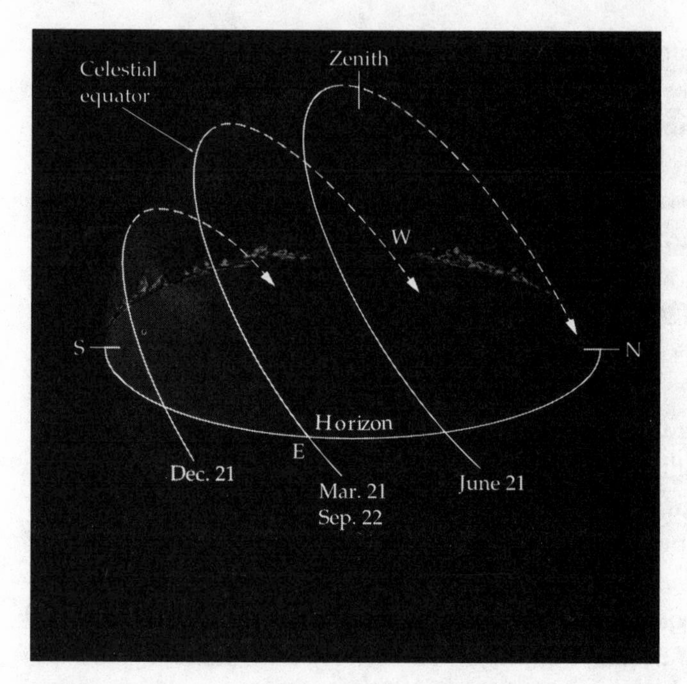

The sun's path in the sky changes with the seasons. In winter, the sun travels in a low arc, rising in the southeast and setting in the southwest. Its maximum southward excursion occurs at the winter solstice. In the summer, the arc is much higher; the sun rises in the northeast and sets in the northwest. Its maximum northward excursion occurs at the summer solstice. Between these extremes, on the first day of spring and the first day of autumn, the sun rises precisely in the east and sets exactly in the west.

horizon, but its cycle is much shorter than the sun's. In fact, the moon completes nearly 13 such cycles in the year that it takes the sun to complete one. The moon also regularly changes its appearance in a separate cycle of phases, from the vanishingly thin crescent of the new moon to the rotund magnificence of the full moon.

Even more curious, a handful of stars appears to shift position relative to the other stars, while maintaining their nightly traverses of the sky. Their courses confined to a broad belt stretched across the sky, these wanderers follow their own peculiar cycles. Close inspection reveals a fine structure of inexplicable whirlpools embedded in their generally smooth and languid motions.

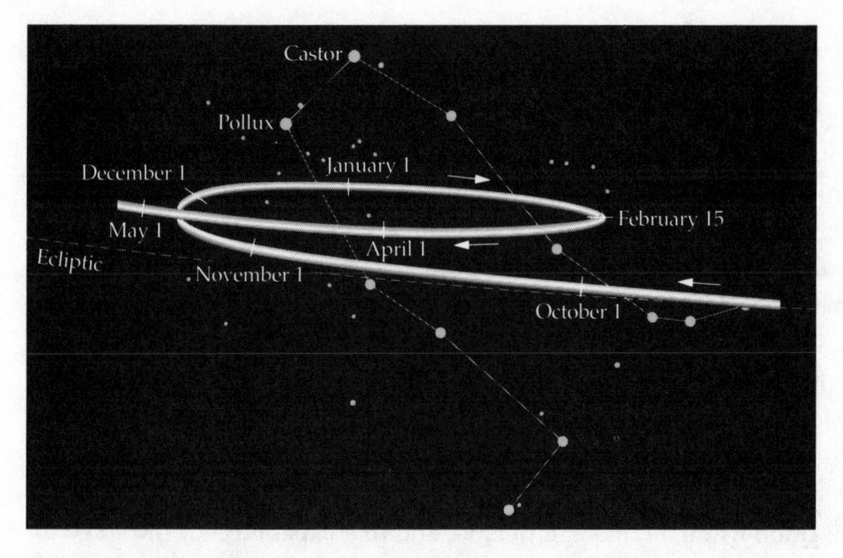

Relative to the background stars, the planet Mars regularly appears to make loops in its path across the sky. For example, when Mars traveled across the constellation of Gemini in 1992 and 1993, it moved during the period from November 30 through February 14 in a direction opposite to that which it normally takes across the sky. The ecliptic (dashed line) marks the sun's apparent path with respect to the stars.

Faced with such hints of complexity, humans naturally began to monitor the movements of lights in the sky. Endeavoring to satisfy an innate curiosity and penchant for solving puzzles, they applied their ingenuity to observing, recording, and measuring celestial events. And they turned to a characteristically human means of expressing the relationships involved and imposing order on them. Beginning with counting and going on to arithmetic and geometry, they had developed mathematics as a practical way to deal with an array of preoccupations, from enumerating livestock, dividing land, and estimating crop yields to toting up wealth, measuring area and volume, and calculating taxes. This wondrous, painstakingly assembled tool kit offered precise techniques for extracting order from confusion and uncertainty—techniques that could also be applied to the cosmos.

Equipped with these mathematical tools, early symbol manipulators soon realized that they could employ geometry to depict celestial motions and arithmetic to express their cyclic nature. By imprisoning their descriptions of the cosmos within a mathematical cage, they believed they could predict and somehow capture what they viewed as the silent but harmonious rhythms of the heavens. There was enough regularity in the movements to suggest that universal laws might govern them, and sufficient complexity and irregularity to spur the development of mathematics and science.

Human needs drove much of this development. Whether it was a matter of defining the seasons, predicting an eclipse, establishing the start of spring planting, fixing the date of a religious festival, or casting a horoscope, astronomical computation supplied reassuring answers. Mathematical investigations into celestial mechanics, which concerns the motions of the solar system, benefited navigation. These studies provided sailors with increasingly precise means of determining their position when out of sight of land, and that capability set the stage for wide-ranging exploration, global commerce, and ceaseless wandering. Those who wished to rule the waves set their best mathematicians to work grinding out tables of future planetary positions and lunar phases. Indeed, that tradition continues in today's astronomical and nautical almanacs, which are produced annually by such national institutions as the United States Naval Observatory and the Royal Greenwich Observatory.

At the heart of all this calculation lies the deeply held conviction that natural phenomena are, in essence, the consequence of just a small number of physical laws, and that these laws are best expressed in the language of mathematics. The goal is to construct a working model of the universe out of commonplace notions: ideas of number and order and measures of time and distance. With such a working model, we can leap ahead in time and predict what the otherwise opaque future has in store for us.

However, mathematics often offers more than one scheme for describing nature. More than 2000 years ago, Greek scholars proposed, pondered, and vigorously debated a variety of ingenious schemes for describing and accounting for the movements of the "wandering stars," which we know as planets. Most put Earth firmly at the center of their theories, but a few took the considerably more daring position that Earth itself was in motion.

Around the year 530, one of the last members of the Academy that Plato had founded nearly a thousand years earlier enunciated the powerful tradition that had been passed down from Plato's time. In his commentary on one of Aristotle's works, Simplicius of Cilicia wrote, "Plato lays down the principle that the heavenly bodies' motion is circular, uniform, and constantly regular. Thereupon he sets mathematicians the following problem: What circular motions, uniform and perfectly regular, are to be admitted as hypotheses so that it might be possible to save the appearances presented by the planets?"

The machinery built to meet Plato's criteria worked remarkably well, and astronomers over the centuries tinkered with its details to improve their predictions. Other than as a point of philosophical or theological dispute, it didn't really matter whether the planets revolved around Earth or around the sun. The question always came down to which set of mathematical formulas could most efficiently generate tables containing the most accurate predictions.

But more than accuracy was at stake. It's much easier to carry around a scrap of paper containing three or four simple formulas than it is to lug around a huge volume filled with facts and detailed observations. The object of a mathematical model, then, is not just to describe natural phenomena accurately, but also to do it in the most compact, economical way possible.

In the seventeenth century, first Johannes Kepler and Galileo Galilei and then Isaac Newton established a physical basis for the mathematical model that best describes natural phenomena. Newton was able not only to extract out of a heap of confusing observational data the underlying physical laws but also to invent a new, labor-saving mathematical tool—the calculus—with which to express those relationships. In essence, he demonstrated that problems in mechanics deal with objects moving in response to forces and that a surprisingly wide array of natural phenomena can be deduced, described, and explained by those forces and by the mathematically expressed laws of motion governing them. Central to Newton's ideas was the notion that the fundamental laws of nature are the same everywhere. A stone dropped from a tower in Italy will behave in the same way as a stone dropped from a tower in England. The moon is subject to the same kind of force as a falling apple.

In his magnificent, revolutionary treatise, *Philosophiae naturalis principia mathematica* (The mathematical principles of natural philosophy), Newton focused on gravity and its effects, but there are hints that he was prepared to go much farther. In the preface, he wrote, "I wish I could derive the rest of the phenomena of Nature by the same kind of reasoning from mechanical principles, for I am induced by many reasons to suspect that they may all depend upon certain forces by which the particles of bodies, by some causes hitherto unknown, are either mutually impelled toward one another and cohere in regular figures, or are repelled and recede from one another. These forces being unknown, philosophers have hitherto attempted the search of Nature in vain."

Newton's equations of motion and his law of universal gravitation were sufficient for handling any problem that involved the movements of bodies in the solar system. With such a fundamental, simple framework in place, it didn't take much of a leap for Newton's successors to imagine planets confined forever to totally predictable, orderly orbits around the sun. Such a perpetual clockwork would never need rewinding or adjusting. This mathematical machinery, by encapsulating completely the solar system's past, present, and future, in principle seemed to leave no room for the unforeseen. It therefore turned astronomers and mathematicians naturally to the question of the solar system's stability.

Concerns about stability arise in practically any dynamical system. Imagine a ball sitting at the bottom of a large, hemispherical bowl. A small nudge forces the ball aside, but it immediately tries to return to its original position. If there is no friction, the ball continues to oscillate back and forth or to trace a curved path along the bowl's contours. This represents a stable situation. In contrast, a ball precariously balanced on top of an overturned bowl represents an unstable situation. A small nudge moves the ball aside, and it continues to move from its original position, never to return.

In the case of the solar system, the question is whether the multifarious influences of every body on every other body shift the planets from their essentially unchanging orbits only slightly and temporarily, or whether these effects can eventually lead to radical, irreversible changes. Billions of years in the future, will the planets continue tracing roughly the same paths as they do now, or will a time come when Mars will catastrophically smash into Earth, or Pluto escape from the solar system? Could Earth itself drift close enough to the sun to become a twin of veiled, noxious Venus?

In one way or another, the problem of the solar system's stability has fascinated and tormented astronomers and mathematicians for more than 200 years. Somewhat to the embarrassment of contemporary experts, it remains one of the most perplexing, unsolved issues in celestial mechanics. Each step toward resolving this and related questions has only exposed additional uncertainties and even deeper mysteries.

The crux of the matter hinges on the fact that it is one thing to write down the equations expressing the laws of motion and a totally different thing to solve those equations. As Newton and his successors quickly discovered, computing the motions of the planets and other bodies in the solar system is no simple matter. In fact, the computations are often so complex that researchers now use supercomputers or specialized, custom-built, electronic machines to solve them.

At first glance, the need for high levels of computing power to describe celestial motions appears puzzling. Apart from asteroids, planetary satellites, and a variety of itinerant but minor bodies, the solar system comprises just nine planets and the sun. The tug of gravity is the only significant force affecting their motions, and the mathematical formula that expresses the relationship between this force and the masses and separations of the planets has been known for more than 300 years. It should be possible to calculate the positions of the planets at any time and to explore what the laws of physics have in store for the solar system. Why isn't it?

If the entire solar system consisted of only the sun and Earth, a straightforward pencil-and-paper exercise would explain all the movements. Just as Newton did three centuries ago, one can use a so-called differential equation to express the physical law governing the moment-by-moment relationships between the positions, velocities, and accelerations of the two bodies. Solving, or integrating, the equation means deducing from these relationships the actual trajectories followed by the bodies in question—where they have been and where they will go. The values of the variables at one instant completely determine those values at all subsequent and preceding times. The mathematics confirms that any orbit in such an idealized, two-body system would be stable. As Newton demonstrated, Earth, for instance, would whirl endlessly around the sun, keeping forever to its elliptical course.

But add another planet to the system, and now there are three bodies tugging on one another. Earth can no longer keep to its precisely elliptical path. It continues to orbit the sun, but, depending on its distance from the other planet, it is affected at different times by a different gravitational pull. Those perturbations distort its trajectory in space, just as Earth's influence in turn perturbs its companion's orbit.

In this case, the equations representing the movements of the three gravitationally interacting bodies spawn no simple mathematical formula that can describe and predict the paths of all three bodies with unlimited accuracy for all time. The problem becomes even worse when four or more bodies are involved. The best that anyone can do is to calculate first the major effects—such as the sun's preponderant influence—and then step by step take into account other, less significant perturbations. Such strings of approximations allow mathematicians and astronomers to close in on the answers they're after. When applied to an actual planetary arrangement, however, this procedure requires horrendous amounts of computation. Indeed, the sheer physical labor involved in applying so-called perturbation theory—the notion that a "real" problem can be solved by making minor modifications to a simple, ideal situation—limits its usefulness as a mathematical tool. This is especially true when one wants to peer ahead billions of years or look back to the origins of the solar system.

Before the advent of reliable mechanical calculators and high-speed electronic computers, the trick was to adopt plausible approximations that provided predictions of planetary positions for a given date with a prescribed level of precision. However, the more precisely one tried to pinpoint a given celestial event, the more calculations it took to make a sufficiently reliable prediction. And the computations were done by hand with the aid of interminable, eye-taxing tables of logarithms. Indeed, until the last half century, the word *computer* meant a person who performed the painstakingly tedious calculations required in astronomy and other number-intensive fields. Whereas present-day astronomers can turn to the electronic machines on their desks, few astronomers in the past could get by without the ranks of anonymous assistants and clerks who performed the necessary calculations.

Electronic digital computers now handle the monotonous calculations required to compile the astronomical tables used in predicting planetary positions and celestial events such as eclipses. But scientists continue to probe the details of those calculations to understand and correct the minuscule defects that mar the accuracy of their predictions. Measurements relayed from spacecraft, along with radar signals bounced off the moon and other objects, enable distance and mass to be determined with an accuracy never before achieved. With the help of this detailed information and the prodigious calculating capacity of computers, scientists can tackle dynamical subtleties that were out of reach just decades ago.

Such precision has many uses. For example, careful computation of both planetary and spacecraft positions, combined with judicious course corrections, steered *Voyager 2* nearly 5 billion kilometers on a 12-year trek from Earth; the spacecraft arrived for its encounter with Neptune on schedule and within a few kilometers of its target. Similarly, historians count on calculations of the timing of eclipses to help date historical events that took place thousands of years ago. Conversely, comparisons between ancient astronomical records and contemporary computations yield discrepancies that shed light on minute changes in the rate at which Earth spins on its axis.

Whether seated beneath a planetarium's darkened dome or before a high-resolution monitor displaying celestial scenes created by sophisticated software, we too can wander freely through time and space, confident that the sky appears to us as it did to Ptolemy nearly 2000 years ago, and as it will to our descendants a hundred years hence.

The quality of observation and calculation is now sufficiently high that scientists can probe the solar system's near future and recent past with a considerable degree of confidence. They can watch orbits evolve over eons and check for signs of instability. They can look into Earth's orbital history for evidence of minute wobbles and changes in orbital shape that may have been strong enough to affect our planet's climate and geological history.

But computation has also clearly demonstrated what mathematician Henri Poincaré at the turn of the century understood but couldn't quite visualize—that mechanics and the laws of physics, as formulated by Isaac Newton, are really much richer than Newton and his disciples

Before the invention of paper, scholars in ancient China recorded significant celestial events, such as total solar eclipses, on fragments of bone and tortoise shell known as oracle bones. This ink rubbing of a piece of tortoise shell bears inscriptions that apparently record the occurrence of a total solar eclipse on June 5, 1302 B.C., as seen in Anyang, China. Researchers used computers to calculate the dates and paths of eclipses visible in China during this period and pinpointed the time of day at which this particular eclipse should have occurred. Because any slowing of Earth's rotation rate would shift an eclipse's path, they could then work out how much the length of a day has increased during the intervening years. (NASA/Jet Propulsion Laboratory.)

dreamed possible. Newton's equations encompass not only the precisely predictable but also the erratic and chaotic. Moreover, this dual nature of the equations appears to mirror the behavior of physical systems that can readily shift from an apparently orderly, predictable type of motion to an irregular, unpredictable course.

The double pendulum is an example of this type of physical system. A single pendulum consisting of a rod pivoted at one end simply swings back and forth. But the addition of a second rod, pivoted from the bottom of the first, greatly increases the motion's complexity. Gyrating like a trapeze artist bent at the waist, the double pendulum oscillates capriciously, sometimes swinging in wide arcs, sometimes pausing for short periods before resuming its unpredictable course.

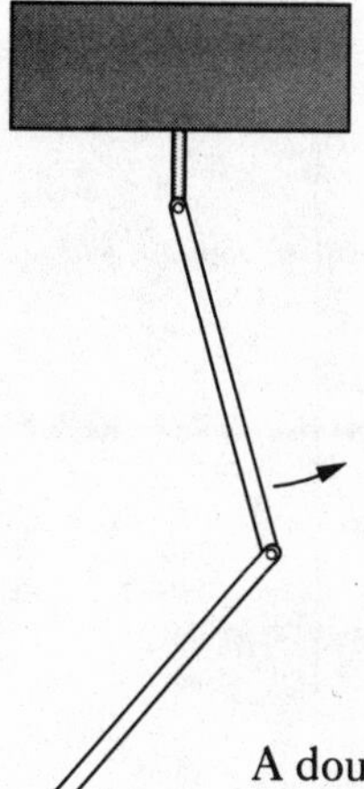

A double pendulum can display the erratic, unpredictable motion typical of a chaotic system.

Sometimes the lower rod does all the swinging, while sometimes the upper rod carries the bulk of the motion. At other times the double pendulum acts as if it were a single pendulum. These remarkable gyrations can have such a strong, hypnotic appeal that museum and airport gift shops regularly offer glitzy contraptions patterned on the double pendulum.

In both science and mathematics, *chaos* is the technical term now used to describe such erratic activity. First applied in 1975 by mathematician James Yorke, chaos refers to the apparently unpredictable behavior of a deterministic system governed by mathematically expressed rules. Although its everyday meaning suggests wild, confused behavior, chaos to its initiates merely signifies limited predictability. The key idea is that the behavior of a chaotic system tends to change drastically in response to a slight change in initial conditions. Two identical double pendulums starting at slightly different positions will very quickly go out of synch and embark on highly individual movement patterns. Similarly, certain mathematical equations will produce very different strings of results if the initial values differ by even a tiny fraction.

Spectacular advances in computing power have brought about a dramatic increase in the understanding of basic dynamics. Taken together, the solutions to the equations that describe a dynamical

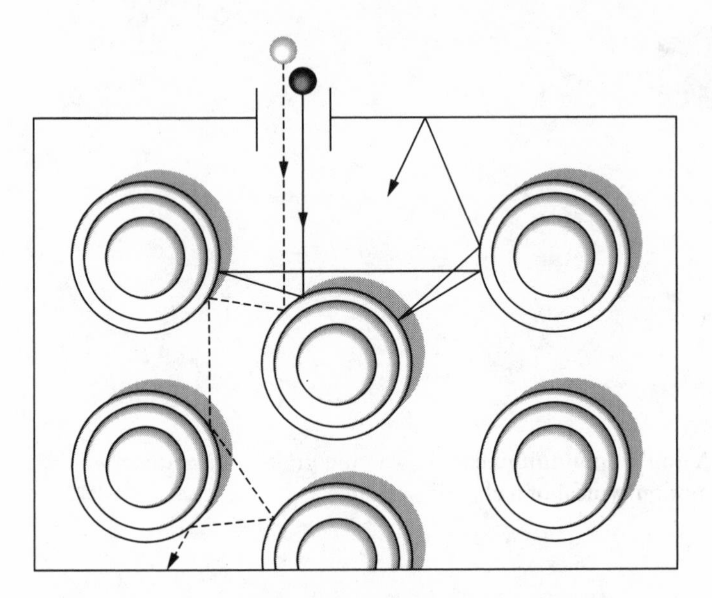

In this physical example of "sensitive dependence on initial conditions," balls with slightly different starting points end up following very different paths when they ricohet through a closely spaced array of cylindrical bumpers on a billiard table.

system typically encompass a variety of coexisting states, some ordered and some disordered. For example, displacing a double pendulum by only a small angle from its vertical starting position produces a regular, predictable motion. But displacing the same pendulum by a large angle initiates an unpredictable, chaotic motion. Order and chaos represent twin manifestations of an underlying determinism. Neither exists in isolation.

Numerical investigations also suggest that the chaos and order of theoretical Newtonian mechanics have counterparts in the solar system. Several strikingly different methods of computing and tracking the evolution of planetary orbits all generate results confirming that chaos lurks in the planetary clockwork. Researchers see evidence of dynamical chaos in the gaps between the orbits of the minor planets of the asteroid belt, in the tumbling motion of Hyperion (one of Saturn's satellites), and in the perplexingly twisted rings that encircle the outer planets. The latest numerical evidence points to traces of chaos in Pluto's peregrinations and even in Earth's orbital motion, although the effects are so far too minute to send either planet careening out of its present orbit. The solar system has apparently survived for more than four billion years in some semblance of its present form, but it isn't quite as placid or predictable as its venerable clockwork image suggests. Nothing guarantees that the future holds no surprises.

However, the presence of chaos in the solar system remains a controversial idea. Not everyone accepts the verdicts emerging from computations that simulate planetary motions over billions of years. Critics argue that approximations are involved in the calculations and that numerical errors inevitably creep in. The fact that previous computational models produced quite different, sometimes contradictory, answers adds to their skepticism.

But computer models of the solar system have improved considerably. Present-day calculations extending millions of years into the past and future are sufficiently accurate that, in the words of modeler Scott Tremaine, "If I were an astrologer, I'm confident that I could tell you what your zodiacal sign was if you had been born a million years ago."

Vastly different methods of computing and tracking the evolution of planetary orbits seem to generate remarkably similar answers. Even more significant is the fact that mathematicians have proved that certain simplified mathematical models of the solar system have chaotic solutions. These achievements are important because, in principle, computation by itself can never prove that a particular system is truly chaotic. However strong the numerical evidence, it can only suggest the possibility of chaos.

None of this work provides evidence that the solar system is falling apart. The real issue is much more fundamental. It's difficult, if not impossible, to make accurate predictions more than a few million years into the future. The dynamics of the solar system contain an element that simply can't be captured by calculation.

If the Newtonian mathematical apparatus used to model the solar system is an appropriate stand-in for the real thing, what does its bizarre, two-faced behavior suggest about what we can determine of the solar system's long-term future? Is there a horizon beyond which no computation can provide a reliable picture? Does this mean the end of our efforts to settle the question of the solar system's stability?

In 1773, at the age of 24, Pierre-Simon de Laplace became one of the first to make a serious attempt at a mathematical proof of the solar system's fundamental stability. Opinion at that time was decidedly mixed on the subject. Isaac Newton believed that divine intervention might occasionally be necessary to put the solar system back in order and prevent its dissolution. Leonhard Euler, impressed by the immense difficulty of accounting for even the moon's motion, despaired of being able to cope sufficiently with the innumerable forces and complicated interactions contained in any realistic model of the solar system. He thought it impossible to make fair predictions of its destiny.

"All the effects of nature are only the mathematical consequences of a small number of immutable laws," Laplace insisted, and he applied his formidable mathematical powers to deciphering the dynamics of the solar system. After completing his exhaustive analysis, he declared that the solar system is stable. The planets repeat their complicated cycles forever, never straying far from their mandated courses.

The notion of Earth and the other planets endlessly whirling around the sun, keeping forever to their nearly elliptical orbits, seems both reassuring and dreary. But Laplace's hard-won proof of such stability applied to a mathematical model of an idealized solar system, not to the real world. His model neglected a variety of subtle gravitational influences that might have changed the outcome of his investigations.

The stability question remains open, but now scientists can explore, to an unprecedented level of detail, related questions about changes in planetary orbits. High-speed computers make possible the creation of sophisticated models of nature, on which experiments can be performed. The result is a new kind of science that is neither true experimentation nor pure theory, and its success depends on the degree to which computational models can be made to behave like natural phenomena. Its capabilities open up areas of research that were inaccessible in the past.

Astronomy, in particular, has felt the impact of the computer revolution. For generations, astronomers working in celestial mechanics were stuck with just one solar system on which to test their theories.

Today, using electronic time-and-space machines, they can in effect audition alternative solar systems and acquire a sense of the variability that nature allows. They can shift the position of a single planet or create a solar system with a unique geometry and then experiment with the system's behavior. And they can do it again and again.

Such experiments may have something to say about the uniqueness of our own solar system. Although its stability hasn't been proved in a mathematical sense, it's evident that the solar system has survived for billions of years in roughly its present configuration. Is this particular distribution of planets just one of many possible stable arrangements, or is it the only one that can produce a stable solar system? Would there be room for another planet, or would its influence disturb the system's apparent stability? Did the solar system in its very early history slough off excess material so that only a stable core remained? To what extent did the need for stability force the solar system into its present arrangement?

In one sense, such questions bring astronomers back to the kind of widely discredited numerology that fascinated scientists like Johannes Kepler. Kepler's first major astronomical work contained an explanation for the existence of precisely six planets—the number known in 1596 when the book was published—at particular distances from the sun. He connected the planetary orbits with the five regular polyhedra, or Platonic figures, by nesting the appropriate polyhedron between each successive pair of spheres defined by the orbits of the planets.

Kepler's ingenious structure wasn't the only one to attach significance to the solar system's numerical parameters. In 1766, Johann D. Titius, a German mathematician, physicist, and astronomer, noted an intriguing, amazingly simple numerical pattern that apparently captured planetary distance relationships. He started with a sequence of numbers: 0, 3, 6, 12, 24, 48, 96, 192, 384, 768, and so on, in which each new term beyond 3 is obtained by doubling the previous term. By adding 4 to each term and then dividing by 10, he obtained a new sequence of numbers, each of which represented the average distance of a known planet from the sun relative to the average distance of Earth from the sun. For the most part, the resulting numbers corresponded closely to the measured ratios.

No one paid much attention to the inconspicuous footnote (appended to his translation of a French science book) in which Titius

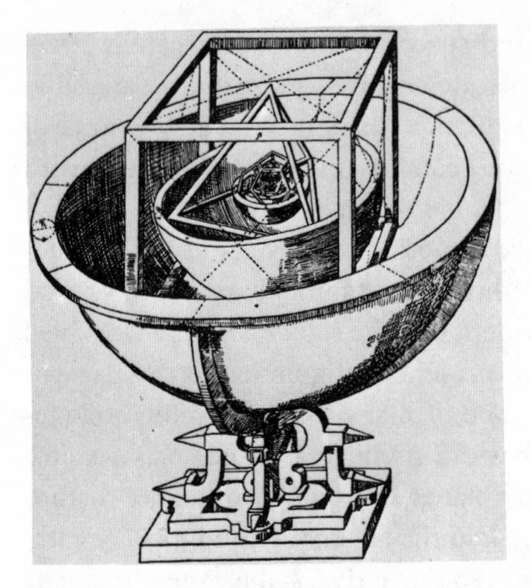

To account for the relative sizes of the orbits of the six planets known in 1596, Johannes Kepler constructed an ingenious nesting of the five types of regular polyhedra, choosing the order giving the best agreement with the known proportions of planetary orbits. A cube separates Saturn from Jupiter, a tetrahedron lies between Jupiter and Mars, a dodecahedron between Mars and Earth, an icosahedron between Earth and Venus, and an octahedron between Venus and Mercury. The sun sits at the system's center. Bowls thick enough to accommodate the eccentric orbits of the planets separate one solid from the next. (Reprinted with permission from David Layzer, *Constructing the Universe.* New York: Scientific American Library, 1984, p. 43.)

first announced this peculiar numerical relationship. Then Johann Eilert Bode of the Berlin Observatory dug it up and made it public in 1772. The Titius-Bode formula gained credibility when the planet Uranus was discovered in an orbit whose average distance from the sun was reasonably close to where the formula predicted it would be. The formula also led to the sighting of the first asteroid, which rendered significant a number in the sequence that previously had been without a corresponding celestial body. But it failed to produce accurate relative distances for the outlying planets, Neptune and Pluto, when those planets were discovered.

The fact that the Titius-Bode formula comes even as close as it does to predicting the positions of the planets continues to intrigue some planetary scientists. Is this numerical relationship really just a coincidence, or does it have something to do with the dynamics and stability of the solar system? Perhaps there was something in the early history of the solar system that inevitably led to such a pattern.

Because we can't experience the five billion or so years remaining before the sun inflates to a red giant and destroys the solar system as we know it, we must rely on mathematics as the most effective and trustworthy means that we have to understand what we see around us. That mysteriously powerful creation of the human mind, which allows us to compress the universe into arrays of symbols on a printed page, continually amazes with its applicability to natural phenomena and with its subtlety and depth.

Distances of the Planets from the Sun:
Actual and as Predicted by Titius-Bode Formula

Planet	*Actual Distance (astronomical units)*	*Distance Applying Titius-Bode Formula (astronomical units)*
Mercury	0.39	$(0+4)/10=0.4$
Venus	0.72	$(3+4)/10=0.7$
Earth	1.00	$(6+4)/10=1.0$
Mars	1.52	$(12+4)/10=1.6$
?	—	$(24+4)/10=2.8$
Jupiter	5.20	$(48+4)/10=5.2$
Saturn	9.54	$(96+4)/10=10.0$
Uranus	19.19	$(192+4)/10=19.6$
Neptune	30.06	$(384+4)/10=38.8$
Pluto	39.53	$(768+4)/10=77.2$

The Titius-Bode formula approximates the relative distances of the planets from the sun. However, the discrepancy between calculated and actual values becomes substantial for the three outer planets (below dashed line), which were discovered after Titius proposed his formula in 1766. One astronomical unit (AU) corresponds to the average distance between the sun and Earth.

A single, tiny planet residing in a solar system dominated by two stars of equal mass would follow an extremely complicated, practically unpredictable orbit. This situation serves as an example of the complex motions possible in a system of only three gravitationally interacting bodies.

The history of astronomy and of the perennial effort to predict the course of celestial bodies and to solve the puzzles of the solar system are intricately intertwined with the history of mathematics and computation. Astronomy was the first mathematical science. It was arithmetic and geometry, in conjunction with observation, that led to the all-encompassing theoretical structure that we have at our disposal today.

It seems in some ways fortunate that our solar system happens not to be so complex that we could never have entertained the notion that simple physical laws govern the universe and have developed a language with which to express those laws. Imagine the tangled motions that astronomers on a planet caught in the clutches of two orbiting stars would observe. Could they ever discern the simple mathematical formulas underlying those erratic movements? Could any form of life, let alone astronomers, even exist in a situation so unstable?

The exploration of our own solar system has taken us from clockwork precision into chaos and complexity. Human limits and computational triumphs marked the halting steps along the way, providing the ingredients of a marvelous mathematical detective story. The tale begins with our distant forebears, who first saw in the night sky a message to pore over and comprehend. It leads to the startling perspectives that the time-and-space machines of modern research have provided of our solar system and the universe in which it drifts.

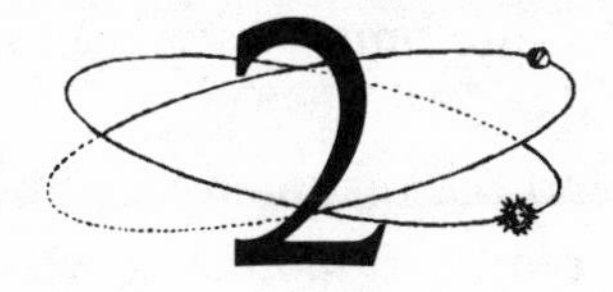

Time Pieces

•

When they come to model Heaven
And calculate the stars, how they will wield
The mighty frame, how build, unbuild, contrive
To save appearances, how gird the sphere
With centric and eccentric scribbled o'er,
Cycle and epicycle, orb in orb.

JOHN MILTON (1608–1674), *Paradise Lost*

SHORTLY BEFORE Easter in the year 1900, two fishing boats chased by a squall anchored off the rocky shore of the island of Antikythera. This barren, craggy islet, barely a mile in length, splits the channel separating the Greek mainland from the island of Crete. It serves as a stark reminder of the perils awaiting any boat that threads the shoals, shifting sandbars, and treacherous currents of this heavily traveled route between the eastern and western Mediterranean.

Six sponge divers and the crews of the two sailing vessels had been on their way home to the small Aegean island of Syme after a hunt for sponges in Tunisian waters. Forced to wait out the storm, the divers decided to spend some time looking for sponges on the rocky shelf where the boats were anchored. When one of them reached the bottom, nearly 200 feet below the surface, he caught sight of the wreck of an ancient ship. An enormous mound of broken marble and bronze statues, tumbled amphoras, and other heavily encrusted, corroded objects lay across its remains. Astonished by his discovery, the diver brought up a larger-than-life bronze arm to prove to his fellow divers that his find was no fantasy.

That bronze arm and the testimony of the fishing boat's skipper persuaded Greek authorities to mount an expedition to recover the treasure. For nine months starting on November 24, 1900, a team of divers painstakingly dug sculptures out of the mud, tied stiff ropes around slimy bodies of bronze and marble, and then returned to the surface as a ship-based crane hauled up the load. Working without air tanks or breathing tubes, the divers could spend only five minutes at a time on the slanting, insecure bottom. The tough work and strain of diving to the required depths limited them to only two dives apiece per day.

The team nevertheless succeeded in recovering several magnificent bronze figures, now displayed at the National Archaeological Museum of Athens. The haul also included both Hellenistic and Roman pottery, a large number of amphoras that may have come from the island of Rhodes, an assortment of marble statues—late copies of early originals—shipped in easy-to-assemble pieces, and a lamp.

At the museum where the objects were taken, staff members focused their attention on the most recognizable fragments, gradually identifying, cleaning, and joining them into more or less complete figures. Periodically, they would examine the various misshapen pieces they had initially set aside, searching for clues to their identity. But the process was slow.

Virtually ignored among the recovered treasures was a dark, porous, compact lump of calcified bronze about the size of a hefty hardbound book. Stored in the open air in a cage outside the museum, the dark lump slowly dried from the outside in. Ancient wood inside the object shriveled to a hard, distorted mass, producing cracks that suddenly separated the lump into four fragments. What the fragments revealed when they were discovered on May 17, 1902, by archaeologist Spyridon Staïs created a sensation. Visible on the surfaces previously hidden within the inscrutable lump were clear traces of toothed wheels, or gears, and extensive, but largely illegible, inscriptions.

Discovered by sponge divers exploring an ancient shipwreck close to the Greek island of Antikythera, this encrusted lump of corroded metal shows traces of an astonishingly sophisticated array of gears for computing celestial movements. (Courtesy of the Adler Planetarium, Chicago, Illinois.)

The gearing was impressively complex. The dials and inscriptions, written in letter forms typical of the first century B.C., suggested an astronomical or calendrical purpose. But all else about this remarkable mechanism remained a puzzle. Indeed, the object seemed too sophisticated to be a product of Greek or Roman technology. Nothing like it had ever been uncovered before.

Initially, scholars speculated that the object was some kind of astronomical measuring instrument, such as an astrolabe, that had originally been mounted in a box. But the mysterious device, which came to be called the Antikythera mechanism, looked far more complicated than any known astrolabe.

Invented during the second or third century B.C. for measuring the altitude of a star or some other celestial body, an astrolabe usually consisted of little more than a circular metal disk hanging from a wooden ring. Suspending the ring vertically from the thumb and using the remaining fingers to steady the disk, the observer moved a pointer pivoted at the disk's center until it was aimed directly at the object under observation. The object's position above the horizon was then read directly from graduated markings on the disk's rim. Many centuries later, navigators used portable, refined forms of the same instrument as their primary means of determining latitude and even the time of day.

Could the Antikythera mechanism have been a kind of navigational instrument, or was it something else? A clock, a planetarium, a sculpture? The controversy over its purpose continued sporadically for

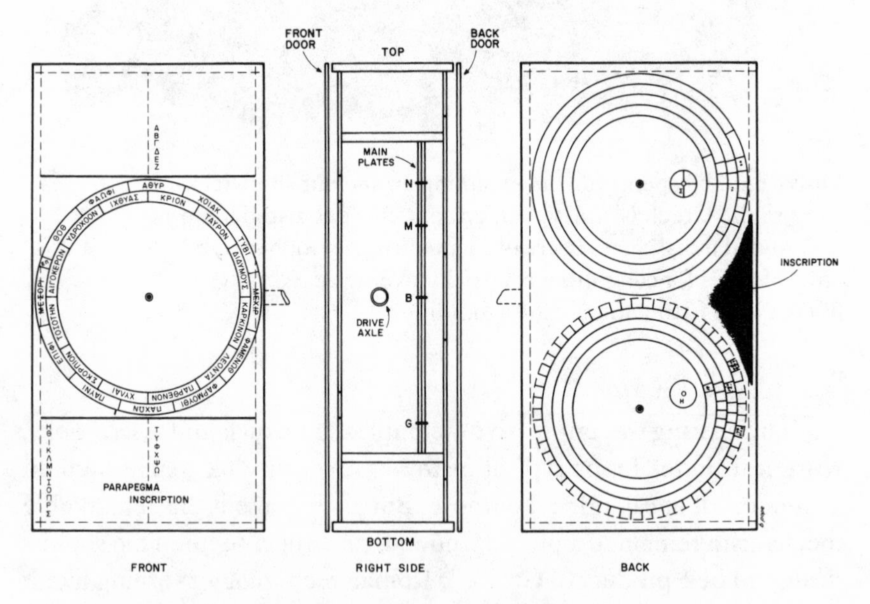

The reconstruction of the Antikythera mechanism reveals a box 16.4 centimeters wide, 4.8 centimeters thick, and 32.6 centimeters tall. Fabricated almost entirely from bronze, the device apparently had front and back doors, front and back plates with dials and inscriptions, and a set of interior plates on which were mounted several arrays of gears. (Courtesy of the Adler Planetarium, Chicago, Illinois.)

decades, while museum technicians patiently cleared away the thick layers of debris that cloaked the bronze fragments.

In 1951, photographs of the fragments caught the attention of Derek J. de Solla Price, a pioneer in the history of scientific instruments. The mechanism's intriguing intricacy brought him to Athens for a firsthand look in 1958, and on the basis of his observations, Price offered a preliminary reconstruction and analysis.

The original object consisted of a box with dials on the outside, both front and back, and a complicated array of gears inside, mounted on a thick bronze plate. Hinged doors protected the dials. Lengthy Greek inscriptions covering all available surfaces described the fabrication and operation of the device. More than anything, it resembled a carefully constructed eighteenth-century clock.

The mechanism was driven by turning an axle that came through the side of the casing. This axle rotated a crown-shaped gear that moved a large, four-spoked wheel. Trains of gears attached to various points on the wheel led to shafts that turned dial pointers, one for the front and two for the back. As the input axle turned, the pointers would all move at various rates around their dials. Signs that the machine had been repaired at least twice—to mend a broken spoke and to replace a missing tooth in a small wheel—suggested that the machine had actually worked.

The front dial had two scales along its rim. An inner, fixed ring displayed the 12 signs of the zodiac. An outer, movable ring showed the 12 months of the year. Both scales were carefully marked off in degrees. The dial's apparent purpose was to show the annual motion of the sun against the background of the stars.

For the inhabitants of the lands surrounding the Mediterranean and for many other peoples throughout the world, the motions of the sun, moon, and stars set the rhythm of everyday life. The sun's daily passage from east to west across the sky marked the days, and a similar motion of stars across the sky marked the nights. In fact, a patient observer of the night sky would notice that the stars slowly circle a particular point high in the sky. They appear fixed to an ebony dome that rotates about the Earth, completing a full circuit every 24 hours. At the same time, the stars maintain their relative positions and form patterns so distinctive and lasting that early astronomers could map and assign names to the groupings, or constellations.

The star trails in this time-exposure photograph of the night sky show how the stars appear to revolve around a point centered on the north celestial pole, which is directly over Earth's North Pole. The stars complete a circuit once every 24 hours. (U.S. Naval Observatory. Reprinted with permission from William J. Kaufmann, *Universe,* 3d ed. New York: W. H. Freeman, 1991, p. 23.)

The starry dome appears tilted because its center of rotation, or celestial pole, isn't directly overhead unless you're at the North Pole. Stars that are positioned high above the western horizon at nightfall follow the sun's example and set during the course of their nightly journeys from east to west across the sky. Other stars rise above the eastern horizon and take their place. Only a band of stars lying sufficiently close to the dome's pole remain visible for the entire night.

This pattern repeats itself over and over again, but not without some subtle changes from night to night. After a few weeks, stars that initially appeared just above the eastern horizon in the evening twilight make their appearance much higher in the sky. Other stars initially low on the western horizon at dusk are no longer visible. They reappear in the dawn sky many months later. Careful observation reveals that in comparison with the rising and setting of the sun and the appearance of twilight, the stars rise and set about four minutes earlier each night. Over the course of a year, that extra time adds up to exactly one day,

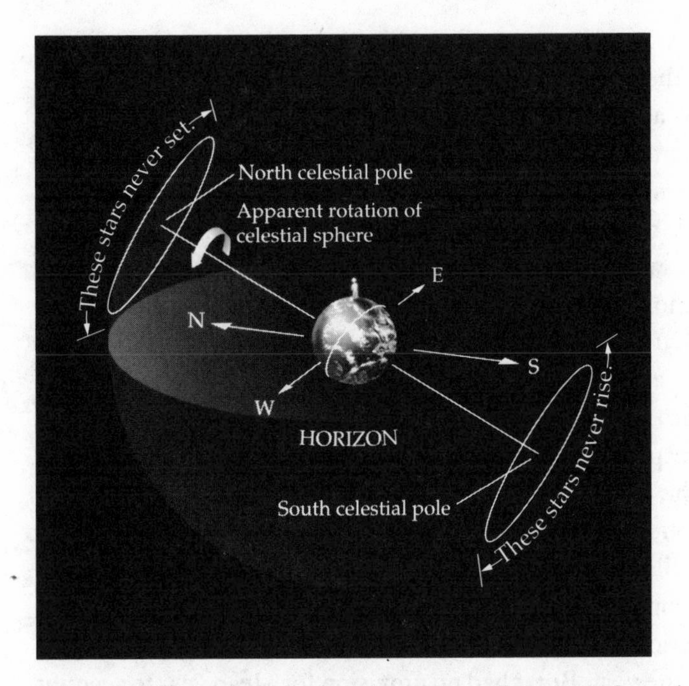

To an observer at any northern latitude on Earth's surface, the sky appears as a hemisphere lying above the circle formed by the horizon. The apparent rotation of the celestial sphere of stars around its axis causes the stars to rise at the eastern horizon and set in the west. Constellations near the north celestial pole revolve about the pole, never rising or setting. Constellations near the south celestial pole lie too far south to be seen from northern latitudes.

and after a full year, the stars that initially make their appearance at dusk are the same as those that were visible at the same time a year before.

The annual cycling of the positions of the stars relative to the sun means that the rising and setting of particular stars and constellations can serve as markers of the seasons. This knowledge was already well-established when the Greek poet Hesiod noted in the eighth century B.C. that spring begins with the rising of the star Arcturus and winter with the setting of the constellation Orion in early twilight.

This pattern also means that the sun's motion differs slightly from that of the stars. It completes its daily circuits marginally faster, thus appearing to travel among the stars from west to east. Its annual path relative to the stars is known as the ecliptic, and the groupings of stars that fall along the ecliptic comprise the constellations of the zodiac. As the sun travels around the ecliptic over the course of a year, it passes through the same constellations of the zodiac at the same season year after year. This yearly track lies not on the celestial equator, an extension of Earth's equator into space, but follows a path relative to the fixed stars, which crosses the celestial equator twice a year at the so-called equinoxes.

The Antikythera mechanism's fixed inner ring had 12 major divisions, one for each sign of the zodiac, with each division further subdivided into 30 parts. Letters inscribed on the zodiac scale appeared to correspond to inscriptions on the plates surrounding the dial, which listed the main risings and settings of bright stars and constellations throughout the year.

The movable outer ring apparently represented the Greco-Egyptian calendar, commonly used by astronomers of that era. This calendar featured 12 periods of 30 days each, followed by an extra block of five days to fill out the year. But it had no provision for a leap year to account for the fact that a year actually lasts longer than 365 days by approximately a quarter of a day. For this reason, the calendar year would step by step drift out of synch with the solar year. Each return of the sun to the same place in the zodiac would take a quarter of a day longer, so the mechanism's outer ring had to be shifted through a quarter of a division to correct for this drift.

Price was able to ascertain that the zodiac and calendar scales were about half a major division out of step. Moreover, from tables of dates of the autumnal equinox (when the sun rises due east and sets due west, and day and night are equally long), historically registered according to the old Egyptian calendar, he could deduce the date for the last position recorded by the mechanism. Assuming that it had remained undisturbed since that time, this particular combination of zodiac and calendar positions could have occurred only in the year 85 B.C., or in some multiple of 120 years (30 days divided by a quarter of a day per year) before or after that year. A short, incised line on the plate near the month scale, half a degree away from the outer ring's final position, possibly served as a reference mark for setting the scale in case it was accidentally shifted. This suggested that the device was made in 87 B.C., was used for roughly two years, and then was taken aboard a ship probably bound for Rome.

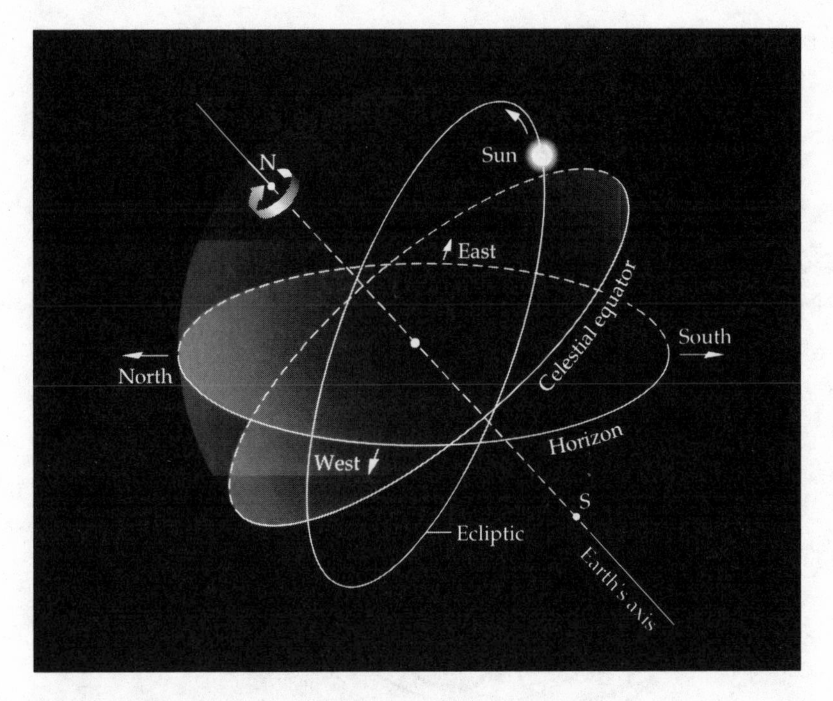

By "freezing" the daily motion of the celestial sphere of stars, it is possible to plot the sun's apparent path relative to the stars over the course of a year. Known as the ecliptic, this track intersects the celestial equator at two points called the equinoxes, meeting it at an angle of 23.5 degrees. Note that although the sun swings from east to west in its daily motion accompanying the stars, it crawls "backward," or from west to east, in its yearly motion along the ecliptic. The sun's annual path as projected onto the celestial sphere runs along the middle of a belt of stars, traditionally broken up into 12 sections, known as the zodiac. The paths of the moon and planets also lie within this belt.

The back dials and plates proved considerably more difficult to decipher. They were both less legible and more complex than the front dial. For instance, the lower dial appeared to have three slip rings, and the upper had four. Each dial also included a small, subsidiary dial resembling the second hand of a clock. Fragments of inscriptions hinted that the dials marked the occurrence of such lunar phenomena as the moon's phases and times of rising and setting.

The mechanism's inner workings remained largely impenetrable until 1971, when Price learned of the possibility of using X rays and

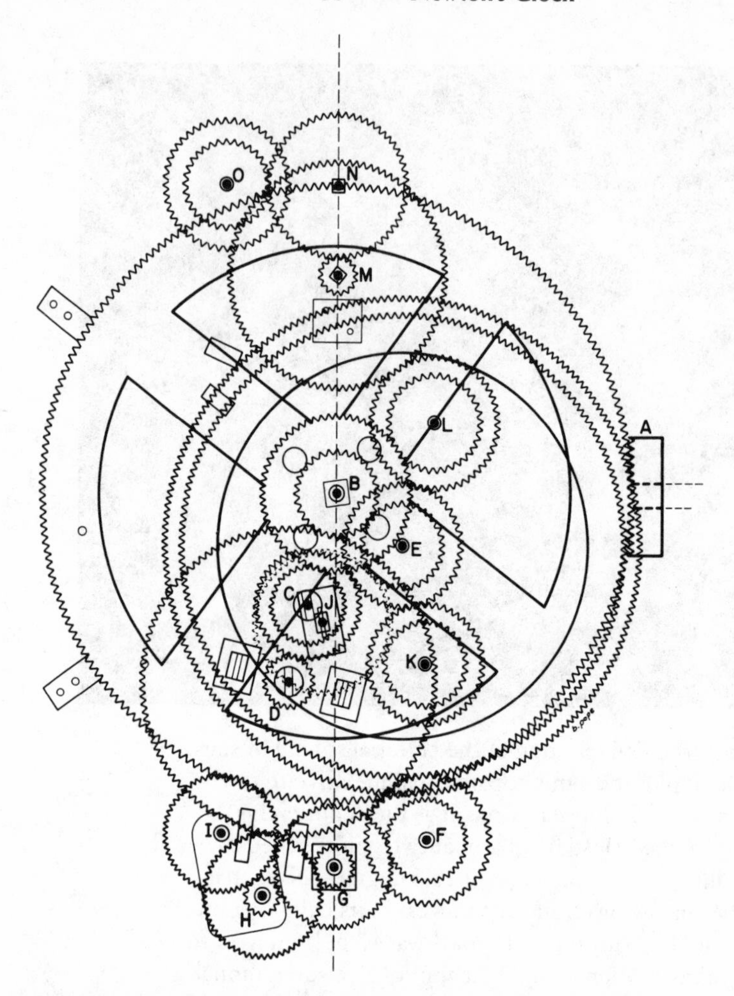

A composite diagram showing all of the gearing in the Antikythera mechanim. (Courtesy of the Adler Planetarium, Chicago, Illinois.)

gamma rays to obtain photographs of its interior. Painstaking work produced a remarkable set of photographic plates that provided sufficient detail for Price and his collaborators in Greece to count the gear teeth and determine much of the structure of the gear trains.

By 1973, it was possible for Price to prepare a preliminary but fairly complete description of the mechanism's gearing. This analysis revealed for the first time that the gear ratios used in the mechanism corresponded to well-known astronomical cycles involving the sun and moon. For example, it employed an arithmetical relationship derived from early astronomy specifying that in 19 years, the moon goes through 235 phase cycles from full moon to crescent moon and back to full moon and passes through the zodiac 254 times.

Whatever the details of its operation, the mechanism was a type of analog computer, using fixed gear ratios to make calculations displayed as pointer readings on a dial. Both its intended purpose and the means used to power it remain purely speculative. Perhaps the device, about the size of a mantelpiece clock, served as a hand-cranked computer for demonstrating cyclic relationships and making astronomical predictions—possibly for astrological purposes—concerning various celestial cycles and noteworthy events observed in the sky. It's also possible to imagine a steady stream of water automatically turning its crank so that the mechanism could operate as an exhibition piece for representing motions in the heavens as they occurred.

Because it mechanized celestial cycles in a numerical fashion, the Antikythera mechanism represented the arithmetical counterpart of the more literal, geometric models previously developed by Greek scholars. Those early models of celestial motions emphasized the geometry of the paths of the sun, moon, and planets across the sky and tried to reproduce those motions geometrically using various combinations of circles. The Antikythera mechanism reproduced not the geometry but the numerical relationships among the observed cycles.

In 75 B.C., not long after the loss of the Antikythera mechanism in a shipwreck, the Roman orator Cicero described an impressive planetarium constructed by Archimedes about 200 years earlier. This celestial globe depicted the motions not only of the sun and moon, but also of the five wandering stars, or planets.

These planets—Mercury, Venus, Mars, Jupiter, and Saturn—are far brighter than any of the fixed stars. Like the sun and moon, they all tend to move from west to east, more or less following the ecliptic. But each planet shows peculiarities in its motion that clearly distinguish it from the others. For instance, Mercury is never more than 22 degrees away from the sun, and Venus no more than 46 degrees. They all traverse the sky at different rates with respect to the fixed stars.

To create his celestial globe, Archimedes did more than simply mark the positions of stars and constellations on a solid ball. He used systems of gears to move various planetary, solar, and lunar markers at their correct rates across a star-sprinkled, hollow sphere. For example, a 30:1 gear ratio linked the annual passage of the sun through the zodiac with the 30 years that it takes Saturn to make the same journey relative to the stars. A similar arrangement generated Jupiter's 12-year period, and so on. Thus, the planetarium designed by Archimedes could reproduce a variety of celestial movements as they occur in the sky from day to day and year to year.

At first glance, it seems curious that, despite the Archimedean precedent, the Antikythera mechanism showed no traces of gearing related to planetary motions. The answer to this mystery lies in several important changes that took place in both astronomical theory and instrument-making between the time of Archimedes and that of the craftsman who made the Antikythera mechanism. The architect of these changes was the mathematician and astronomer Hipparchus, who lived during those intervening years and probably worked on the island of Rhodes.

The most important change in thinking concerned planetary motions. Although the planets travel eastward most of the time, they also appear to change course periodically. At various times, they reverse their motion and start traveling westward, eventually coming to a stop and reversing direction again to resume an eastward course. As a result, they trace a loop relative to the stars.

To account for the looping, or retrograde, paths of the planets, Hipparchus refined the notion that this apparent motion is really a consequence of regular movements along a combination of two circles: a larger circle, or deferent, with Earth at its center, and a smaller circle, or epicycle, whose center travels along the larger circle. The planet itself is tied to the epicycle, which rotates as its center moves along the larger circle.

This theoretical complication made the use of simple gearing to represent planetary motions practically impossible for craftsmen of Hipparchus's era. Indeed, as far as we know, no further mechanical models of planetary motion appeared until the medieval period in Europe.

A meticulous, patient observer, Hipparchus also discovered a tiny but significant creep in the positions of the fixed stars. At the spring equinox, midway between winter and summer, the sun is at a particular, well-defined place in the zodiac, and it returns to that spot every

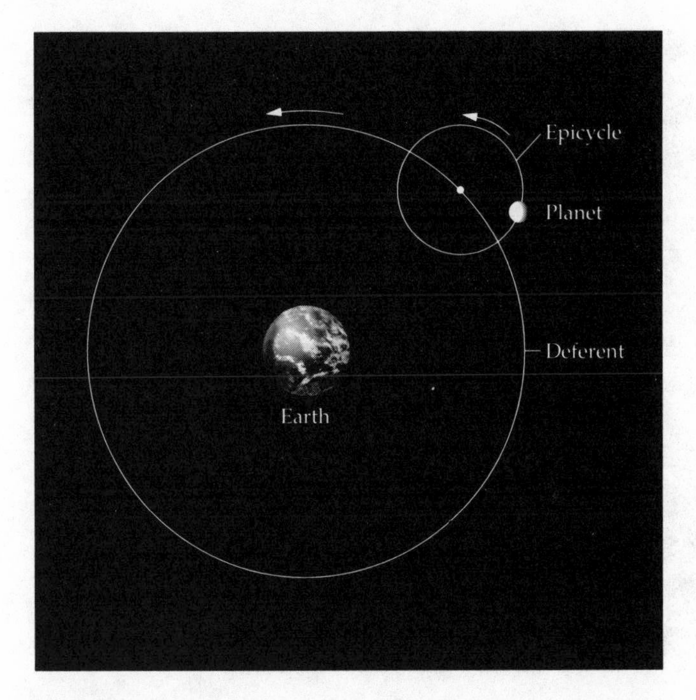

To explain the occasional loops in the apparent motion of the planets in the sky, astronomers in ancient times imagined that each planet travels with a uniform motion around a small circle, known as an epicycle. The epicycle's center, in turn, moves on a bigger circle called the deferent, with Earth in the middle.

year. By carefully measuring the longitudes of stars (that is, their positions along the ecliptic starting from the spring equinox, where the celestial equator cuts the ecliptic), Hipparchus noticed that his star positions differed slightly from those recorded a few years before. If the interval between observations was more than a century, the difference was more than 1 degree. In other words, the sun didn't return each year to exactly the same patch of stars in the zodiac.

Hipparchus interpreted this discrepancy as the slow slipping of the zodiacal girdle around the celestial sphere, which came to be called the "precession" of the equinoxes. Precession refers to the excruciatingly slow shift in equinox positions westward along the ecliptic. Because this is the reverse of the sun's motion along the ecliptic, the equinoxes occur a little earlier with each cycle and gradually move through the zodiac. This shift, though minuscule from day to day, is significant for

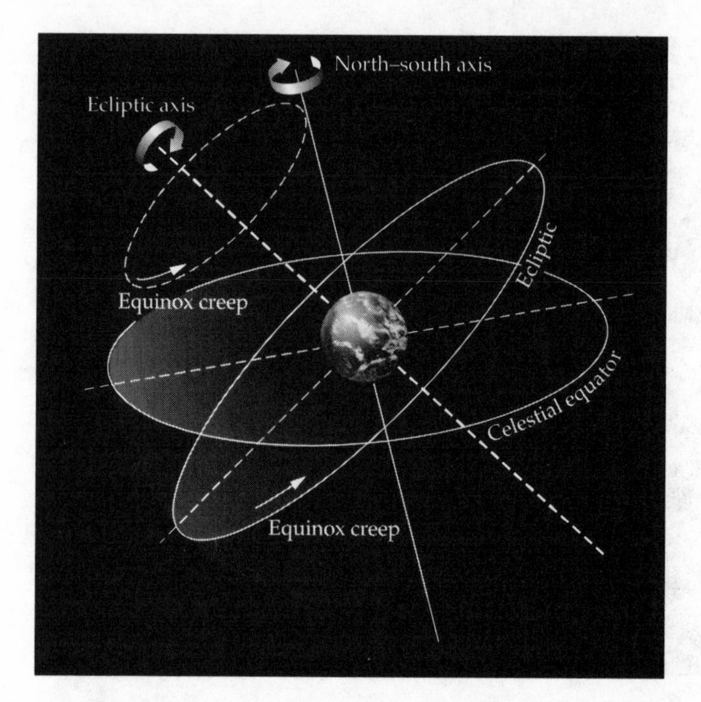

In addition to the daily motion of the whole heavens around the north–south axis fixed to Earth and the yearly motion of the sun along the ecliptic in the zodiac band of stars, Hipparchus discovered a slow rotation of the whole pattern of stars around a different axis at right angles to the ecliptic. The resulting apparent motion very slowly shifts the position of the vernal equinox, where the celestial equator and ecliptic intersect, westward. This slow drift is known as the precession of the equinoxes.

astronomical measurements, and astronomers have allowed for it since the time of Hipparchus.

Hipparchus contributed one other element that influenced the design of the Antikythera mechanism. He introduced a mathematical procedure, known as stereographic projection, that enables a sphere to be mapped onto a flat surface. A simple formula gives every point on the sphere a particular spot on the planar map. This distorts the sphere's geometry but maintains the relative positions of objects. The whole process is roughly equivalent to shining a light through a transparent sphere marked with points and lines to create a shadow on a piece of paper.

The transition from a solid, three-dimensional representation to a flat, two-dimensional surface enabled makers of models to avoid the cumbersome and inconvenient use of a globe. It suggested the possibility of depicting the universe on flat plates and reproducing its movements on dials. Indeed, the astrolabe, which began as a means of measuring the altitude of a celestial body, gradually evolved into a much more elaborate device, with networks of arcs inscribed on its solid disk, inscriptions depicting zodiacal signs and the ecliptic, and a

The astrolabe gradually developed from a simple device for determining the altitudes of stars into a refined navigational instrument and an elaborate display piece replete with astronomical lore. This English astrolabe dates from the fifteenth century. (Smithsonian Institution.)

variety of indicators marking the positions of the dozen or so brightest stars.

The discoveries of Hipparchus led to a shift in emphasis away from the five planets to displays of the cycles of the sun and moon. Planetary motions were shown in simplified form or accounted for by appropriate inscriptions rather than complicated epicyclic gear assemblies. Thus, the Archimedean planetarium evolved into just the sort of solar and lunar calculator represented by the Antikythera mechanism.

Price's study of the Antikythera fragments led him to conjecture that the mechanism originated on the island of Rhodes. The device may have been the product of a craftsman associated with the school of Posidonios, a prominent Stoic philosopher and astronomer of the period. Although no longer as prosperous and powerful as it had been in earlier centuries, Rhodes still flourished as a high-technology center, producing a variety of sophisticated crafts and manufactures. An occasional ally of rapidly expanding Rome, Rhodes also attracted a number of notable Romans to its schools. Cicero himself visited the school of Posidonios in 78 B.C. Given his great interest in mechanical devices and astronomy, it's not so farfetched to suppose that he may have seen the mechanism there. Indeed, it's not impossible that the ship wrecked off the island of Antikythera was on its way to Rome, carrying Cicero's souvenirs of his travels in the Greek islands.

But this is all speculation, and the evidence is fragmentary. "I must confess," Price wrote, "that many times in the course of these investigations I have awakened in the night and wondered whether there was some way round the evidence of the texts, the epigraphy, the style of construction and the astronomical content, all of which point very firmly to the first century B.C."

Taken at face value, the Antikythera mechanism—the sole survivor of what was undoubtedly a long tradition of astronomical automata—served primarily as an elegant simulation of the heavens. It was a miniature monument to Greek and Alexandrian astronomy. Although nobody seriously believed that a device such as the Antikythera mechanism or the planetarium of Archimedes proved that the stars and planets were turned by gears in a kind of celestial clockwork, these models played a significant role in the development of a mechanistic philosophy for explaining how nature worked, from the sublime motions of heavenly bodies to the intricate coursing of bodily fluids.

These ingenious devices also illuminated the intimate link between mathematics and astronomy, especially the role of number in astronomical prediction. By demonstrating an ability to predict the movements of the moon, the rising and setting times of stars, and the changes of the seasons, astronomers could please their rulers while contemplating the evident mathematical order of the heavens.

About 200 years after the construction of the Antikythera mechanism, Claudius Ptolemaeus, better known as Ptolemy, put down on paper an equally sophisticated, but theoretical rather than mechanical, apparatus for making astronomical and astrological predictions. More a synthesizer than an original thinker, Ptolemy compiled the works of the Greek astronomers, especially Hipparchus, into a masterly exposition that dominated Greek, Arabic, and medieval thought for the next 14 centuries. His writings marked the triumph of a new, mathematical attitude toward geometrical models. In Ptolemy's words, "Our problem is to demonstrate, in the case of the five planets as in the case of the sun and moon, all their apparent irregularities are produced by means of regular and circular motions (for these are strangers to disparities and disorders)."

In his most influential and massive work, which he called the *Syntaxis mathematicae* (The mathematical collection) but which more often goes by its Arabic designation, the *Almagest,* Ptolemy propounded a theory that described the motions of the sun, the moon, and the planets. Employing simple rules but allowing for complex details, he prescribed a set of mathematical procedures that he claimed would faithfully mimic the relevant motions of the solar system. Indeed, like a set of crank-driven mechanical gears, Ptolemy's system of displaced (eccentric) circles and epicycles, with carefully adjusted radii, tilts, speeds, and displacements, reproduced the heavenly movements with reasonable accuracy. Anyone using his mathematical machinery could quite efficiently grind out reliable predictions of planetary positions. Though lacking the true elegance and clear focus of a unified, coherent system, his method furnished a remarkably comprehensive collection of tools for deciphering the universe.

Ptolemy's compelling, easily visualized, Earth-centered geometric model became the central authority in astronomy. Medieval astronomers generally accepted both his data and his methods, patiently deriving new predictions of the positions of the planets,

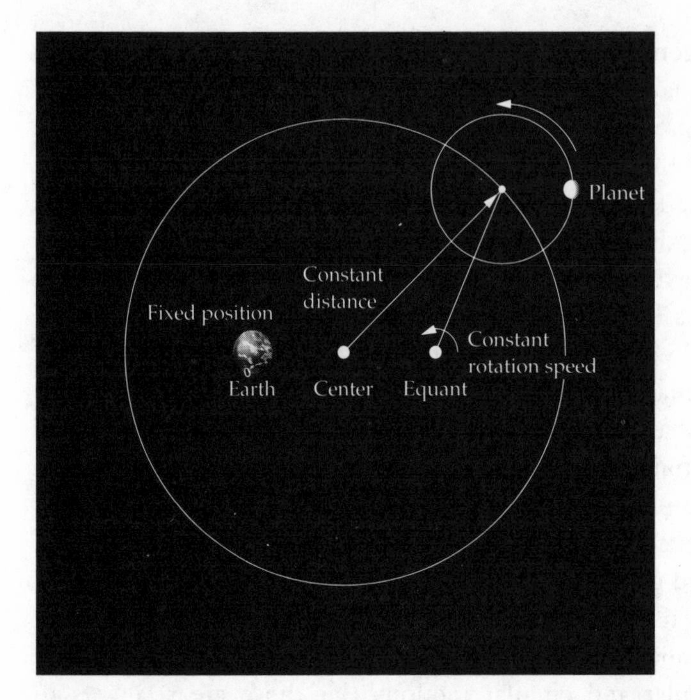

Ptolemy's Earth-centered scheme of planetary orbits required the use of both epicycles and equants (imaginary points about which the centers of epicycles would rotate uniformly). In addition, Earth itself was no longer in the middle of the circle about which an epicycle's center revolved. By selecting suitable radii and speeds of rotation, Ptolemy could reproduce the apparent motions of the planets with remarkable accuracy.

the times of the rising and setting of the moon, and the dates of lunar eclipses and other celestial events. In most cases, however, the predictions were good for only a few years at best, and new values had to be computed at regular intervals. Furthermore, over the centuries, discrepancies between the calendar and the occurrence of certain key celestial events, such as the spring equinox and the full moon, became unduly large, causing considerable consternation among religious authorities, who determined the dates of celebrations such as Easter on the basis of these events. Compiling the necessary tables and ensuring that the figures matched observed phenomena became a sophisticated art, essential for chronology, astrology, and navigation.

By the time Christopher Columbus sailed westward from Palos, Spain, in 1492, navigators were already using hefty printed volumes containing astronomical tables and instructions for manipulating navigational instruments and computing geographical positions from celestial observations. Columbus himself probably carried copies of two invaluable books. The "perpetual almanac" prepared by Abraham Zacuto, who had taught at Salamanca University earlier in the century, contained more than 300 pages of astronomical tables that had already contributed to such navigational feats as Vasco da Gama's famous expedition from Portugal around the tip of Africa to India. The second volume, called the *Ephemerides*, had been produced by the prominent German astronomer and mathematician Johannes Müller, who went by the Latin name Regiomontanus.

At the time of Columbus's first voyage, printing with movable type was less than 50 years old, but book publishing had already become a vibrant, thriving enterprise, which contributed to a crucial flow of information throughout Europe. In the same year, 19-year-old Niklas Koppernigk (Copernicus) was studying astronomy and its close kin, astrology. His initial sojourn at the University of Cracow was the start of a long road that led to medical and legal studies in Italy and eventually to the revolutionary, counterintuitive concept of a sun-centered solar system.

The astronomical tables that Columbus consulted during his voyages proved useful for determining latitude and, to some extent, longitude. A prediction contained in the tables probably also saved his life at a crucial moment during his fourth voyage to the lands he had discovered.

Nearly two years after sailing from Cadiz in 1502, Columbus and his restless, disgruntled crew were stranded on the north coast of Jamaica, confined to worm-eaten, leaking ships. The native inhabitants of the region were no longer awed by the newcomers. Annoyed by their voracious appetites and angry at the depredations of crew members who had plundered several villages, the population was hostile and would no longer supply food. Weary, arthritic, and prematurely aging, Columbus withdrew to his ship and pondered his precarious situation. Returning to the stained pages of the *Ephemerides*, he noted Regiomontanus's prediction of a total eclipse of the moon early in the evening of February 29 of that year.

1497	1500	1501
Finſter der Sunnē	Finſter des mōdes	Finſter des mōdes
29 3 1	5 14 2	2 17 49
Des Hewmōdes	Des Wintermōdes	Des Maien
Halbe werung	Halbe werung	Halbe werung
0 35	1 37	1 52
Drei punct	Czehen punct	

1502	1502	1504
Finſter der Sunnē	Finſter des mōdes	Finſter des mōdes
30 19 45	15 12 20	29 13 36
Des Herbſtmōdes	Des Weinmōdes	Des Hornungs
Halbe werung	Halbe werung	Halbe werung
1 7	1 1	1 46
Czehen punct	Drei punct	

A page from the *Calendarium* of Regiomontanus shows his prediction of the lunar eclipse of February 29, 1504. (Courtesy of Owen Gingerich.)

Such an eclipse occurs only when the moon passes into Earth's shadow. A lunar eclipse looks the same from anywhere on Earth, but it occurs at different times, as measured by local clocks. Regiomontanus's book contained not only the expected dates of eclipses but also diagrams illustrating how completely the moon would be covered and precise information about each eclipse's duration and timing down to the hour.

Columbus had observed a lunar eclipse on an earlier voyage and had noticed discrepancies between the predictions made by Zacuto and those contained in the *Ephemerides*. Moreover, he had no reliable way of determining the correct local time of this particular projected eclipse. The times provided by Regiomontanus for its start and end were for Nuremberg, Germany. Despite these uncertainties, Columbus was desperate enough to take a chance. On the day before the predicted eclipse, he summoned the leaders of the native inhabitants and warned them through an interpreter that if they did not cooperate with him, the moon would disappear from the sky on the following night.

The natives for the most part were unimpressed; some even laughed. Columbus nervously awaited the outcome of his gamble. Could he rely on tables that had been compiled several decades earlier and that predicted the positions of celestial bodies only for the years between 1475 and 1506? How large were the errors? Had there been a misprint?

Amazingly, the prediction proved correct. As the full moon rose in the east on the appointed night, Earth's shadow was already biting into its face. As the moon rose higher, the shadow became larger and more distinct until it completely obscured the moon, leaving nothing but a faint red disk in the sky. The natives were sufficiently frightened by this unexpected apparition and by Columbus's uncanny prediction to beg for forgiveness and appeal to him to restore their moon to the sky. Columbus responded that he wished to consult with his deity. He retired to his quarters, using a half-hour sandglass to time how long the eclipse would last. Some time later, when the eclipse had reached totality, he emerged to announce that the moon, in answer to his prayers, would gradually return to its normal brightness.

The next day, the natives brought food and did all they could to please Columbus and his crew. Columbus himself used the timing of the eclipse to calculate his ship's longitude, but his answer proved wildly erroneous. On June 29, 1504, a Spanish ship rescued Columbus's stranded party, a year after it had first beached on the Jamaican coast. A few months later, Columbus set sail for Spain, bringing to an end his voyages to the New World.

Both Regiomontanus and Columbus lived in an era of rapid change, when many traditional ideas came under challenge. The explorations of adventurous seamen brought enormous new wealth, creating an elite class with the leisure and funds to support a wide range of artistic and scholarly endeavors. These voyages also highlighted the huge discrepancies between the claims of classical geographers and what the navigators had actually found. Obvious errors in Ptolemy's geography made it easier to question his celestial system as well.

Astronomy stood at the center of much of practical life in the fifteenth and sixteenth centuries. Emperors, kings, and even town governments often employed mathematicians as advisers schooled in the intricacies of astrological prediction. These hired savants not only taught the sons of wealthy citizens and nobles but also compiled almanacs for political and religious leaders and for ordinary citizens. They provided prognostications ranging from weather forecasts to predictions of natural catastrophes, plagues, and political and economic prospects. Indeed, it was a time of considerable religious anxiety and upheaval, punctuated by apocalyptic fears of an impending, climactic struggle against the forces of evil. Such fears were fanned by the growth of movements to reform what many, including Martin Luther, saw as a hopelessly corrupt, misguided church.

Those tumultuous times also saw the first glimmerings of a revolution that would transform the art of astronomical prediction, culminating in a fundamentally different, more fruitful mathematics tied to the notion of cause and effect in the physical world. Müller, or Regiomontanus, whose book aided Columbus in his travels, stands as a figure caught unknowingly in the midst of this great transition. Regiomontanus had learned Greek to avoid having to read Ptolemy's works in faulty Arabic translations. He produced a seminal volume containing the essence of these astronomical writings along with a commentary, bringing to the attention of the Renaissance world works that had been lost and later rediscovered, only to be distorted by careless or unwitting translators. Copernicus consulted this book and subsequent, unabridged translations of Ptolemy's *Almagest* in his own studies.

In 1475, just before his death at the age of 40, Regiomontanus also prepared a manuscript that presented the mathematical problems of finding the true position of a comet. He provided techniques for determining a comet's distance from the sun, its diameter, and the length of its tail. Regiomontanus didn't directly challenge the long-standing notion that comets, which mysteriously appeared and faded away, were not celestial and could therefore exist only in the region between Earth and its moon. But he assumed that, like the stars, they could be studied mathematically. Regiomontanus's modest treatise, finally published in 1531, provided one of the steps along the lengthy, winding path that brought not only recurring events but also such transient ones as the appearance of comets into the mathematical purview of astronomers.

As this new confidence in the power of the intellect to grasp the universe and harness it to human purposes grew, the stage was set for the brilliant achievements of Copernicus, Kepler, and Galileo.

Wanderers of the Sky

•

Man has weav'd out a net,
and this net throwne
Upon the heavens,
and now they are his owne

JOHN DONNE (1573–1631),
Ignatius

THE IMPOSING manor house on the Benatky estate 35 kilometers northeast of Prague stands on a hill overlooking the floodplain of the river Jizera. It was here, at the end of the sixteenth century, that Tycho Brahe had settled to build a great astronomical observatory matching the one on the Danish island of Hven, which he had established but been forced to leave for political reasons. Intrigues at court had led to cutbacks in the grants he customarily received to operate his observatory and research center, and his relationship with a newly enthroned king of Denmark had deteriorated rapidly.

For Johannes Kepler, the bone-shaking, six-hour coach trip from Prague to Benatky on February 4, 1600, marked a turning point in his life. He dreamed of finding a serene, secure temple of scholarship where he could pursue his vision of a harmonious universe created according to a rational plan that could be read in the stars and planets. From Tycho, he hoped to extract the observational data he needed to advance his study of the fundamental harmonies of the cosmos.

A year before he met Tycho, the young Kepler already had a clear vision of what he wanted to achieve. In a letter dated February 16, 1599, to his former professor at the University of Tübingen, he spoke enviously of Tycho's jealously guarded treasury of data: "My opinion of Tycho is this: he is superlatively rich, but he knows not how to make proper use of it as is the case with most rich people. Therefore, one must try to wrest his riches from him."

What Kepler found was a household in noisy turmoil, crowded with laborers, craftsmen, visitors, servants, and Tycho's own family, along with his astronomical assistants and retainers. Tycho, ensconced as a favorite of the superstitious, star-struck ruler of the Holy Roman Empire, was caught up in costly renovations of the commodious, three-story house at Benatky. With its clear views of the horizon in all directions, Tycho sought to convert the house into a suitably grand observatory-in-exile. Kepler, fresh from the relative obscurity of the provincial town of Graz in what is today Austria, must have felt out of place. His long-anticipated encounter with Tycho surely had some awkward moments.

When they first met, Kepler was 28 years old and Tycho was 53. By that time, astronomy no longer took center stage in Tycho's life. Already widely recognized for his meticulous astronomical observations, the aging and homesick, but still imperious, astronomer was pondering more than ever his place in history. He had become extremely sensitive to people's opinion. Kepler, on the other hand, brimmed with audacious, ingenious ideas. He never seemed to hesitate to think the unthinkable and question the unquestioned. Astronomy was his particular passion.

The two men also presented a strong physical contrast. The aristocratic Tycho, accustomed to bending people to his will, towered over the lowly and insecure Kepler, who in his letters and other writings often described himself as "doglike" and worse.

Tycho's chief distraction was a bitter feud with Nicolai Reymers Ursus, his erratic predecessor as imperial mathematician, who lived up to his self-conferred last name, which means "bear" in Latin. The

Left: Tycho Brahe (1546–1601). *Right:* Johannes Kepler (1571–1630). (Smithsonian Institution.)

dispute centered on Tycho's unshakable belief that Ursus had stolen from him the idea for a model of the solar system in which the sun circled a stationary Earth, with the remaining planets orbiting the sun. This unwieldy plan represented Tycho's attempt to reconcile the venerable, Earth-centered Ptolemaic solar system and the controversial, sun-centered version introduced by Copernicus. Inordinately proud of his scheme but characteristically cautious, Tycho had decided to keep it secret until he could obtain additional observational data to support it.

The trouble started when Ursus, four years after visiting Tycho at Hven in 1584, published an astronomical tome containing a model of the solar system strikingly similar to Tycho's own. Believing that Ursus had discovered the idea in manuscripts stored at Hven, and infuriated by his audacity, Tycho accused Ursus of plagiarism. The offense had continued to rankle Tycho, and 12 years later, Kepler's fortuitous arrival suggested to Tycho a means of exacting a measure of revenge.

Kepler's entanglement in this messy affair arose from the desire of a young, exuberant scholar, having published an important first work, to impress his colleagues and elders. Kepler had made his debut with the *Mysterium cosmographicum* (The secret of the universe), completed in 1596. To Kepler, the planets sweeping around the sun along their predestined paths carried a hidden message. He sought in the ratios of the different planetary distances from the sun a pattern that would explain the heavens and unveil the mind of their Creator. He firmly believed there was nothing capricious or arbitrary about either the number of planets or their distances. The full title of Kepler's book expressed his ambitious goal: "A Forerunner to Cosmological Treatises, containing the Cosmic Mystery of the admirable proportions between the Heavenly Orbits and the true and proper reasons for their Numbers, Magnitudes, and Periodic Motions."

Many months of exploratory calculations eventually led to a geometrical model (described in Chapter 1) in which the five regular polyhedra lay neatly between the spheres of the six planets. Because only five such regular polyhedra exist, this nested structure fixed both the number of planets and the relative sizes of their orbits. Although the ratios in Kepler's tidy scheme didn't quite match the measured distances, he was no doubt pleased that his rigid geometry would admit of six and only six planets.

Kepler had put a great deal of effort into his book, and he was proud of his achievement. He sent copies of the volume, sometimes accompanied by somewhat fawning letters, to such well-known scholars as Tycho and Ursus. Through a quirk of fate, two copies ended up in the hands of Galileo Galilei, a virtually unknown, unpublished professor who was nine years Kepler's senior.

Kepler had given the copies to an ambassador about to depart for Italy, with instructions to present them to anyone who might find the material of interest. Forgetting his obligation until just before leaving Rome, the ambassador hastily cast about for potential recipients and settled on Galileo, who quickly scribbled a note of thanks, which was eventually conveyed to Kepler.

For Ursus, Kepler's package arrived at an opportune moment. In the midst of preparing a virulent personal attack on Tycho, he saw immediately that he could use Kepler's testimonials to his scholarly stature to bolster his own reputation at Tycho's expense. When the resulting volume, *De astronomicis hypothesibus* (On astronomical hy-

DE COMETA ANNI 1577. 97

NOVA MUNDANI SYSTEMATIS HYPOTYPOSIS AB AUTHORE NUPER AD-inventa, qua tum vetus illa Ptolemaica redundantia & inconcinnitas, tum etiam recens Copernicana in motu Terræ Physica absurditas, excluduntur, omniaq; Apparentiis Cœlestibus aptissimè correspondent.

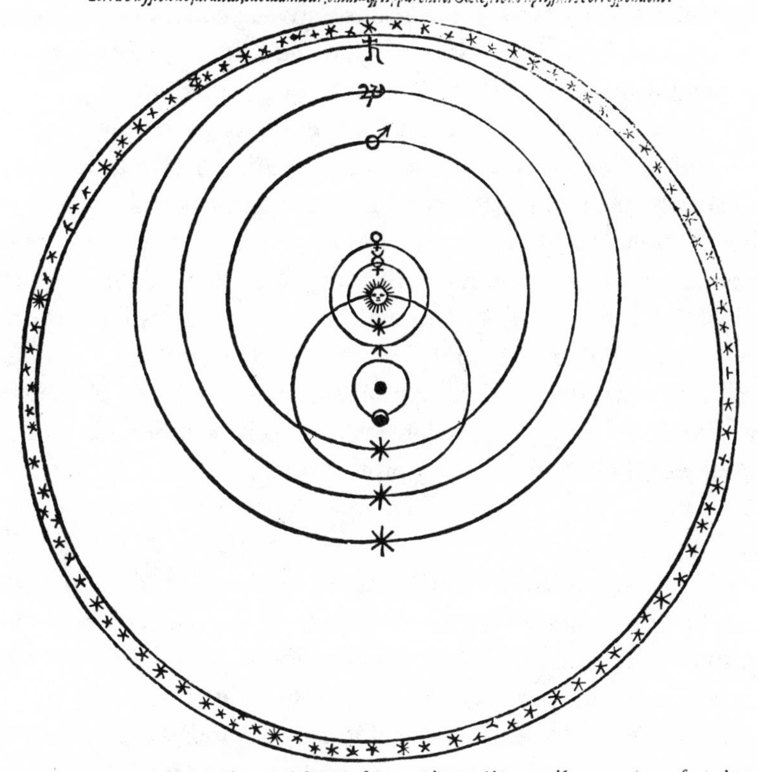

Pleniorem verò hujus novæ Orbium Cœlestiū dispositionis explicationē, inter quædā magna totius præsentis elucubrationis corollaria, circa finē Operis addere cōstitui; ubi per Cometarū motus prius ostensum & liquidò comprobatū fuerit, ipsā Cœli machinā non esse durū & imperviū corpus varijs orbibus realibus confertū, ut hactenus à plerisq; creditū est, sed liquidissimū & simplicissimū, circuitibusq; Planetarū liberis, & absq; ullarū realiū Sphærarū opera aut circūvectione, juxta divinitus inditā Scientiā administratis, ubiq; pareré, nihil prorsus obstaculi suggerere. Vnde etiā constabit, nullā absurditatem in hac Orbiū Cœlestiū ordinatione ex eo sequi, quòd Mars Acronichus Terris propior fiat, quā ipse Sol. Neq; n. Orbiū aliqua realis & incōgrua penetratio (cū illi revera Cœlo nō insint, sed docēdi & intelligendi rē gratia saltē proponātur) hoc modo admittitur, neq; ipsa ullorū Planetarū corpora sibi unquā occurrere possunt, aut motuū Harmoniā, quā singuli eorū observant, ulla ratione interturbare, utut Mercurii, Veneris & Martis imaginarij Orbes Solari permisceantur, eundemque transeat, prout hæc latiùs eo in loco, circa totius (ut dixi) Operis Colophonem; præsertim verò in volumine nostro Astronomico, ubi ex professo de his agemus, apertius declarabitur. Nunc

To reconcile Ptolemaic and Copernican concepts of the cosmos, Tycho envisioned a solar system in which the sun and moon move around a fixed Earth, with the remaining planets revolving around the sun. (Library of Congress.)

potheses) appeared in 1597, it contained the text of Kepler's innocently obsequious letter, which Tycho was bound to notice.

When Tycho and Kepler finally met, after a lengthy comedy of stray letters and crossed signals, Tycho, in turn, was planning to use the younger man to repudiate Ursus. Tycho desperately needed an expert to bring order to his observational data on the planets. But Kepler's real value to a bear-hunting, obsessive Tycho lay not so much in his mathematical expertise as in the damage Kepler's defense of his system could cause to Ursus's already tattered reputation.

Tycho's dream of a splendid observatory safely removed from court intrigues evaporated when Emperor Rudolph II recalled him to the capital for ongoing astrological consultations as the political situation in Bohemia deteriorated. Kepler, too, had much to complain about during his months at Benatky and then in Prague. He was subject to a host of humiliations that disturbed his composure and aroused his volatile temper. He was saddled with the vexing, pointless task of writing an argument in support of Tycho's planetary system. He faced the daily jealousies of Tycho's other assistants. He lacked an officially recognized professional position and contract. And he had to endure Tycho's reluctance to part with certain pieces of observational data.

Quarrels between Kepler and Tycho erupted again and again, only to be patched up or smoothed over for a time. Kepler bitterly noted in a letter that Tycho gave him "no opportunity to share in his experiences. He would only, in the course of a meal and in between conversing about other matters, mention, as if in passing, today the figure for the apogee of one planet, tomorrow the nodes of another."

In the end, however, economic necessity forced Kepler to stay and accept what crumbs he could scrounge from Tycho's table. A political upheaval in Graz in the summer of 1600, aimed at eradicating Protestantism, foreclosed any thought Kepler may have had of returning there. As a Lutheran in a strongly Catholic region, he had lost both his property and his official position as schoolteacher and provincial mathematician.

Yet even in this turbulent period, Kepler showed the characteristic energy and doggedness that were to mark all his ventures. Despite the demands of business matters and several months spent in Graz in an attempt to recover at least a portion of his family's property, he plunged into a variety of studies, including a series of investigations that culminated in important advances in the understanding of optics.

Kepler's status remained uncertain for some time. He had no official position and no salary other than what Tycho could offer from his own funds. Finally, in August of 1601, Tycho took Kepler to meet Emperor Rudolph II. That meeting resulted in a deal in which Tycho agreed to publish his planetary theories, which provided techniques for accurately calculating planetary positions, in a volume to be named for and dedicated to the emperor. Kepler gained recognition as Tycho's official assistant.

Tycho died on October 24, 1601, repeating over and over, "Let me not seem to have lived in vain." Kepler described the scene in his diary, in which he had carefully recorded important events in Tycho's household. About two weeks later, he was appointed Tycho's successor as imperial mathematician.

Those who held this august position were expected to supply meteorological forecasts, cast horoscopes, and make other predictions as needed by the emperor. Earlier in the year, to improve his prospects and perhaps in anticipation of Tycho's death, Kepler had prepared a calendar containing his prognostications for 1602. The resulting book, *De fundamentis astrologiae certioribus* (On the more certain foundations of astrology), presented not only Kepler's predictions but also an attempt to reform what he believed was a valuable classical doctrine, just as Martin Luther had reformed Christianity less than a century before.

Kepler believed that the physical properties of the sun and planets have important effects on Earth's weather, but he firmly rejected the notion, held by most other astrologers, that the division of the stars along the ecliptic into zodiacal signs had any meaning. He expressed both perplexity and faith in a letter he wrote in 1599: "How does the face of the sky affect the character of a man at the moment of his birth? It affects the human being as long as he lives in no other way than the knots which the peasant haphazardly puts around the pumpkin. They do not make the pumpkin grow but decide its shape. So does the sky; it does not give the human being morals, happiness, children, fortune, and wife, but it shapes everything in which the human being is engaged."

Kepler's interest in establishing a physical basis for astrological predictions prompted him to consider scientific phenomena that few had previously thought to ponder. For example, from the assumption that the sun provides whatever light and warmth Earth experiences, Kepler concluded that the influx of solar radiation at Earth's surface could be computed and its summer and winter values compared. Although the mathematical tools available were clearly inadequate for anything but a crude calculation, Kepler's musings were insightful and profound. He even discussed how the accumulation of heat in the atmosphere causes the period of greatest warmth to follow the summer solstice (that is, the time when the sun is highest in the sky) rather than coincide with it.

Although Kepler's salary remained uncertain and irregular because he lacked the guile necessary to press his case on an unstable ruler nearing bankruptcy, his position was ensured. Kepler was now able to focus on untangling the knotted motion of Mars, a problem with which he had failed to make as much progress as expected when Tycho was alive.

The subtleties of Mars's motion provided a crucial test of any mathematical theory purporting to capture planetary peregrinations. Like the other planets, Mars wanders within a band along the ecliptic, generally moving from west to east while oscillating to and fro across its mean trajectory. It sometimes even changes direction and moves backward for a time before resuming its eastward course. But Mars appears to deviate more than the other outer planets from a circular orbit around the sun.

Past astronomers had labored to create a mathematical apparatus that would crank out reasonable approximations of this wayward planet's future position and so provide adequate guidance for astrologers and navigators. They had looked for a convenient way to describe these motions mathematically, without worrying greatly about the physical reality of their geometrical constructions or the true nature of what they observed in the sky. In Ptolemy's geometrical recipes, for example, it made no difference whether the points of light seen in the night sky were patches of glowing gas or massive bodies. There was no reason to ask what causes Mars and the other planets to follow the courses they do. A description of their wanderings was sufficient, and that didn't depend at all on understanding why.

It was obvious to astronomers from earliest times that planets don't follow simple paths across the sky or travel at uniform speeds. Because there was no evidence whatsoever that Earth itself moves, they naturally assumed that Earth is stationary. That also avoided the complication of having to interpret observations made from a moving platform. And past astronomers also assumed that planetary motions could be described by combinations of circles.

But circles weren't enough to account for the planets completely. Because they appear to move, and in fact do move, at different speeds at different times, and because the sun's progress along the ecliptic also undergoes a seasonal variation, astronomers introduced the idea of the equant: an imaginary point, some distance from Earth, about which speeds look uniform. Thus, an observer situated at the equant would see the planet moving at the same speed at all times, even though its distance from the equant varied. Indeed, Ptolemy's geometrical apparatus for describing planetary motions included both equants and epicycles (see Chapter 2). Though his scheme was laid out carefully enough to supply reasonably accurate predictions of planetary positions (all one had to do was follow the steps and use the tables Ptolemy had drawn up), it was also intricate enough that considerable effort and a great deal of time were required to perform the necessary calculations.

Fourteen centuries later, Copernicus again took a strictly descriptive course but chose a different frame of reference for his mathematical apparatus. By shifting the sun to the center, he hoped to simplify the computational equipment needed to make predictions, but he found he still required circles moving on circles. His plan for the planets had the additional disadvantage of suggesting that Earth is in motion, which seemed to contradict everyday experience.

Contrary to the idea held by many people that both the Ptolemaic and Copernican schemes were thickly encrusted with epicycles upon epicycles, neither model was as complicated as it sounds. Ptolemy used just one large epicycle for Mars. Although Copernicus needed more circles in his sun-centered model than Ptolemy did in his Earth-centered scheme, Copernicus's total of 34 circles for all the planets and the moon didn't represent a great proliferation.

The Copernican model was an elegant one, and it sparked keen debate and wide-ranging speculation among the scholarly elite during the decades following its creator's death in 1543. The appeal of this counterintuitive plan of the solar system lay partly in the way it provided natural explanations for celestial phenomena that had been recognized but had never been explained. By setting a spinning Earth free to move around the sun, it made sense of the precession of the equinoxes as an additional motion of Earth—a slow drift in the direction of its spin axis. What had once appeared arbitrary now seemed more reasonable.

However, the choice of mathematical model was still only a matter of taste or philosophical predilection. In the absence of physics from the equations, the mathematical model could take whatever form fitted the observations. Indeed, with sufficient ingenuity, both an Earth-centered and a sun-centered scheme could be modified indefinitely to accommodate new observations and generate reliable predictions of planetary motions. As historian Owen Gingerich has pointed out, "These two cosmologies were in effect geometrical transformations that produced identical predictions, so that merely transforming to a sun-centered system was insufficient by itself to produce better tables. By the same token, errors in prediction could at least initially be corrected within a geocentric framework as easily as in a sun-centered one."

Kepler broke this impasse by expounding and championing a new basis for selecting a mathematical model. He was convinced by the elegance of the Copernican scheme and imbued with the same conviction of a natural order that drove Pythagoras and his followers in ancient Greece to search for an underlying numerical harmony. Kepler maintained that the physical universe was laid out according to a mathematical design that was simple and accessible to human intelligence. The motions of the planets appeared discordant, he argued, because no one had yet learned to hear their songs. The natural philosopher's task was to identify the prime cause from which all else logically followed.

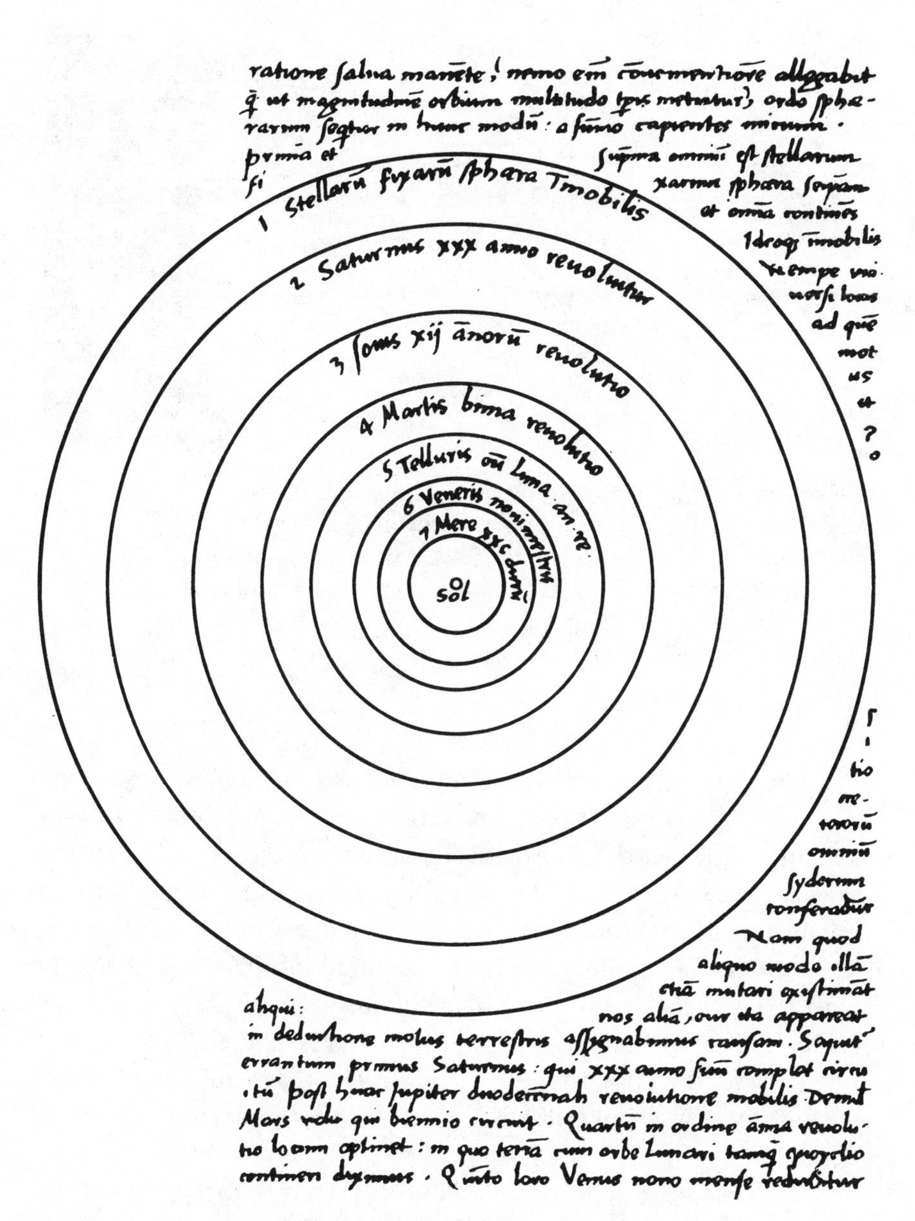

In Copernicus's own drawing of the solar system, the planets, including Earth, orbit the sun, and only the moon revolves around Earth. (Reprinted with permission from Joseph Silk, *The Big Bang*, rev. and updated ed. New York: W. H. Freeman, 1989, p. 19.)

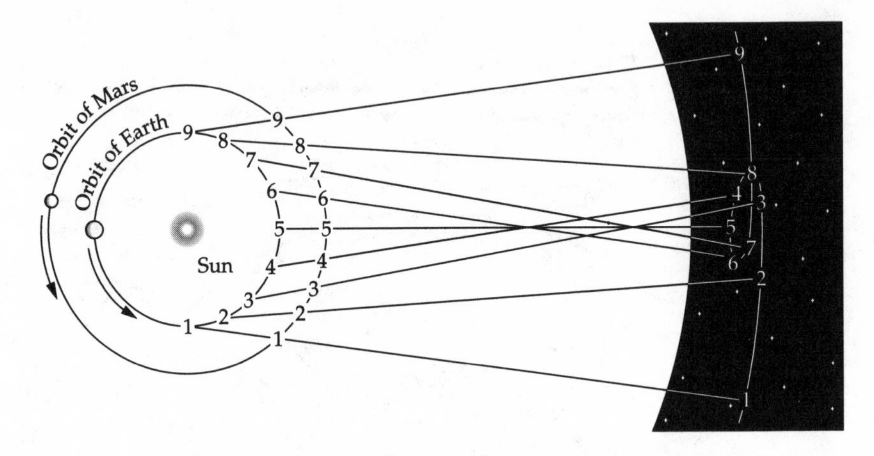

A sun-centered model of the solar system explains the occurrence of loops in the apparent paths of the planets in terms of the relative speeds of the planets. Earth travels around the sun more rapidly than does Mars. Thus, as Earth overtakes and passes the slower moving planet, Mars appears (from points 4 to 6) to move backward for a few months.

These ideas had already formed the basis for Kepler's first major work, the *Mysterium cosmographicum*. In that volume, he developed his key argument that the physical body of the sun causes the planets to move as they do. In Kepler's view, the sun itself had to play a major role in any theory of the planets. How the sun exerted that influence, however, remained ill-defined. But he found inspiration in analogies with the way the intensity of light diminishes with distance from a source and with the way magnetism behaves, a topic of particular interest with the publication in the same year that Kepler first arrived at Benatky of William Gilbert's influential text *De magnete* (On magnetism).

Kepler came to believe that a force emanating from the sun sweeps the planets along in their paths, like a paddle wheel stirring up water flecked with debris. This concept guided his interminable numerical investigations, but the physical laws Kepler devised and continually modified were fundamentally flawed. He remained caught up in the Aristotelian notion that there is no motion without force, even as Galileo was slowly and secretively developing the concept of inertia.

Galileo's studies suggested that once a body is set in motion, it continues indefinitely in a straight line unless some force acts on it. In contrast, Kepler envisioned the dance of the planets as the result of a titanic struggle between the sun's influence and the planets' own natural resistance to motion.

By insisting on a physical reason for motion, Kepler was able to reject the idea of imaginary points about which planets appear to move in uniform circles at constant speeds. The instantaneous calculations a planet would have to make, Kepler argued, to continue describing a uniform circle from one moment to the next in the absence of a compelling physical force, were impossible. How could a planet know when to slow down or speed up? Out of these considerations came the first glimmering of the modern notion of an orbit as the trajectory a body follows when acted on by a physical force. The solid and crystal spheres envisioned by ancient astronomers and philosophers to carry the planets around the sun did not exist. At the same time, Kepler insisted, the sun and the planets themselves were no mere points of light but enormous physical bodies seen from a great distance.

When Kepler undertook to decipher the orbit of Mars and develop a mathematical model, or planetary theory, to fit Tycho's observations, he had to distinguish that planet's motion from the orbit of Earth. He was in the position of an observer noting the coordinates in the sky of a planet whose orbit is unknown, while riding around the sun on a planet whose orbit is likewise unknown. To work out a precise numerical relationship between orbital speed and distance from the sun, Kepler faced the great challenge of choosing appropriate geometries for the two orbits so that a line joining Mars and Earth projected to the stars would always correctly mark the position of Mars relative to the fixed stars as seen from Earth.

What makes this feasible is that orbits repeat themselves at regular intervals. For example, astronomers can determine the position of Mars on the same date in the calendar for several years in succession (with a small correction for the fact that Earth takes about a quarter of a day longer than 365 days to complete its circuit of the sun). In effect, Earth takes a year to return to exactly the same spot in its orbit, but because Mars travels more slowly, its position as viewed annually from

Earth differs from year to year. From such observations carried out over a long enough period, it would be possible to plot the orbit of Mars. Similarly, astronomers could deduce the shape of Earth's orbit by using the fact, well known to Copernicus, that Mars takes 687 days to travel around the sun. By collecting and plotting observations of Mars separated by 687 days, they would obtain the same data as they would by looking from a stationary Mars toward a moving Earth.

With the benefit of hindsight, Kepler's task sounds straightforward. In reality, he faced immense conceptual, observational, and mathematical obstacles.

Kepler owed a great deal to Tycho's obsession with the precise measurement of celestial positions. Using custom-designed instruments, Tycho and his assistants measured planetary and stellar positions so meticulously that their naked-eye observations rarely

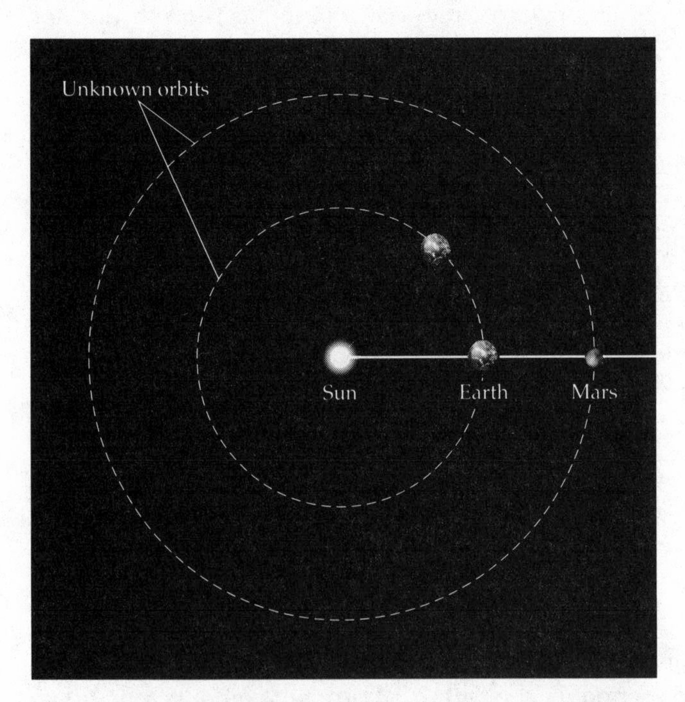

To find the shape of Earth's orbit, Kepler used the fact that Mars returns to the same point in its orbit every 687 days while Earth, traveling more quickly, occupies a different position when Mars returns.

erred by as much as 2 or 3 minutes of arc, about a tenth of the diameter of the moon's face as seen in the sky. Tycho's real contribution, however, which Kepler used to great advantage, was the realization that astronomy needed precise and continuous data. It was useful to determine planetary positions not just at crucial alignments but also at various other points along an orbit. Indeed, Tycho's habit of repeating his observations set him apart from his predecessors and nearly all his contemporaries, who were largely content with much less precise measurements and who generally bothered to make only one or two observations at key moments in the celestial drama.

Tycho's store of data on Mars, amassed over 30 years, was particularly accurate and complete. Aware of its value, Kepler made sure that he retained control of the notebooks after Tycho's death, despite concerted efforts by fractious members of Tycho's family to reclaim them.

Nonetheless, the precision of the raw data was guaranteed only to the extent that Tycho's crews of youthful apprentices correctly handled the cumbersome, unfamiliar instruments in his observatory. Much also depended on how well Tycho and his assistants massaged the numbers.

Astronomers were aware, for example, of the distorting effects caused by refraction, whereby the atmosphere bends the path of light so that objects near the horizon tend to appear higher in the sky than they really are. Determined to ensure the greatest possible precision, Tycho worked out tables and rules for estimating how much to add or subtract from individual measurements so that refraction and other optical effects (such as parallax, which is the change in the apparent position of celestial bodies viewed from different places at different times) were taken into account. Kepler himself, despite his myopia and limited experience in making astronomical observations, was a keen student of the behavior of light; even while struggling with the problems of Mars, he published a major book on optics in 1604, which included several important discoveries.

In addition, in order to compare and map planetary motions, a planet's position, determined by measuring how far away and at what angle a planet lay from a reference star, had to be mathematically transformed into angles—a longitude and a latitude—with respect to the ecliptic. The mathematical machinery for handling these calculations was already quite well established by the time Kepler began his work. The principal tool required for this task was trigonometry, which specified a variety of formulas, given certain lines and angles, for computing the lengths and values of other lines and angles. Sky calculations required spherical trigonometry, a mathematics tailored to geometric figures on the surface of a sphere.

To unravel Mars, however, Kepler needed much more than trigonometry; moreover, the mathematics available to him—mainly arithmetic and Euclidean geometry—could be applied to problems involving distances that changed over time in only the most tedious, circuitous way. Computation with Arabic numerals had begun in Europe during the thirteenth century, but 300 years later, handling fractions and using division were still considered difficult operations. Decimal fractions, taken for granted in modern-day life, didn't appear until 1585. Kepler himself generally worked with the ratios of large whole numbers to express relationships for which we now automatically use decimal fractions. Trigonometric computations—basic chores that today's student, using a simple electronic calculator, can perform in seconds—required hours of computation by teams of human calculators. Tycho usually had a good supply of these assistants, but Kepler often lacked the funds to hire such help and had to perform the calculations himself, repeatedly adding, subtracting, multiplying, and dividing interminable sets of whole numbers.

Because of the importance of determining how orbits evolved over time, Kepler also had to develop schemes by which the static geometry of the early mathematicians could take into account the dimension of time and the changing geometries inherent in planetary motion could be approximated. He struggled, for instance, to determine the areas of unusual or irregularly shaped geometric figures. Nowadays, such problems are easily solved using the calculus or computer-based methods to arrive at numerical approximations. Kepler had to rely on geometrical techniques that involved dividing the figure into a host of smaller pieces, computing their areas, and then adding the individual areas together.

Other considerations further complicated Kepler's quest. To devise his theory of the orbital motion of Mars and then to apply that mathematical apparatus to the remaining planets, Kepler had to choose

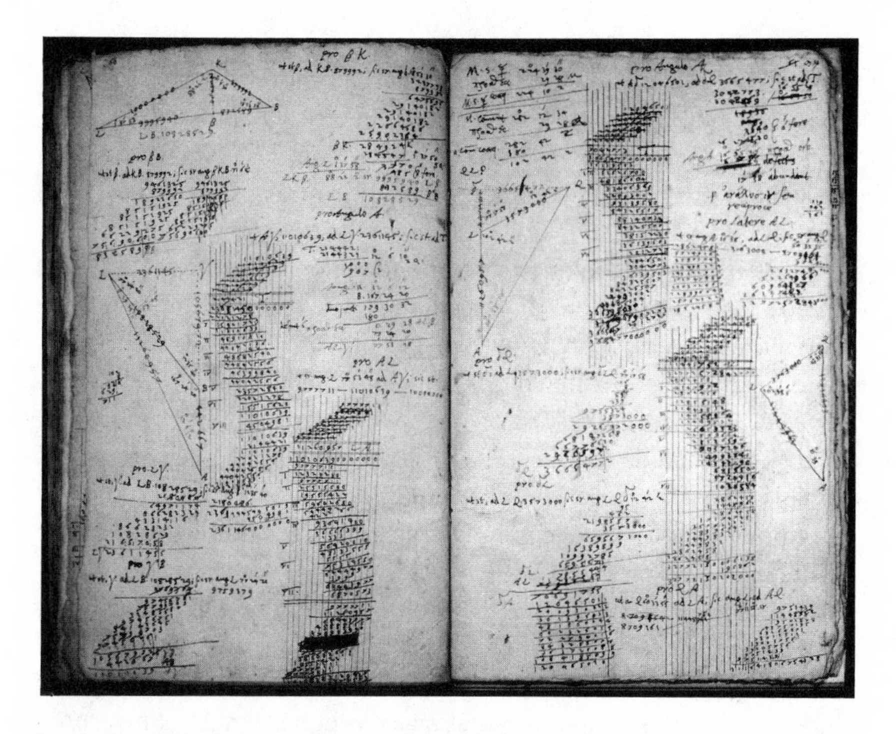

A sample of Kepler's notations and calculations. (Courtesy of Owen Gingerich.)

the data that would determine how well his concepts were confirmed by observation. He had to select the appropriate model, and he had to decide how far from the chosen data his theory could wander without threatening its validity. Where the theory apparently failed, he had to determine what changes in his mathematical apparatus would improve its correspondence with the data. Overall, Kepler's procedure had the same quality of trial and error common today when researchers use high-powered computers to simulate physical phenomena and compare successive iterations against observation. Indeed, Kepler probably would have felt quite comfortable at a modern supercomputing center.

Kepler outlined his aims and what he believed he could achieve in a letter he wrote on February 10, 1605: "My aim is to show that the heavenly machine is not a kind of divine, living being, but a kind of clockwork . . . , insofar as nearly all the manifold motions are caused by a most simple, magnetic, and material force, just as all motions of the clock are caused by a simple weight. And I show how these physical causes are to be given numerical and geometrical expression."

Kepler's prodigious effort to bring harmony to planetary motion culminated in a large, tall book published in 1609. Although by 1605 he had worked out the key elements required to account for the positions of Mars, Kepler had to spend several more years battling Tycho's heirs, who were aghast at the idea of contributing to the Copernican heresy, to obtain their permission to publish his book. Then he had to find the funds and locate a printer to publish it. *Astronomia nova, aitiologetos seu physica coelestis* (A new astronomy, causally explained, or celestial physics) not only solved the Mars problem but also presented a sustained argument rejecting the old descriptive astronomy of Ptolemy and Copernicus in favor of a new astronomy based on physics.

Buried within its pages, this volume contained the two statements that we now call Kepler's first and second laws. Planets travel around the sun in elliptical orbits with the sun at one focus of the ellipse, and they sweep out equal areas in equal times, as measured by the area covered over a particular period by an imaginary line joining the planet to the sun. In other words, a planet moves more slowly along its elliptical orbit when it is far from the sun than it does when it is near. These conclusions represented a radical renunciation of much of celestial mechanics as it was known and practiced in Kepler's time. They downgraded the circle to a special case and put the sun not only at the geometrical but also at the physical center of the solar system.

Kepler's treatise was a bold attack on the predominance of circles in astronomical thinking. Aristotle had taught that "what is eternal is circular, and what is circular is eternal." Circular motion was the only perfect and natural motion, and centuries of dogma favoring this interpretation had lifted the circle to the status of a sacred object. But Kepler wasn't entirely alone in breaking with tradition. In the previous century, Renaissance artists studying the laws of perspective could readily imagine ellipses as circles seen from an angle. Even earlier, in architecture, graceful elliptical arches started to supplant the more ponderous circular arches of Roman and medieval bridges and other structures.

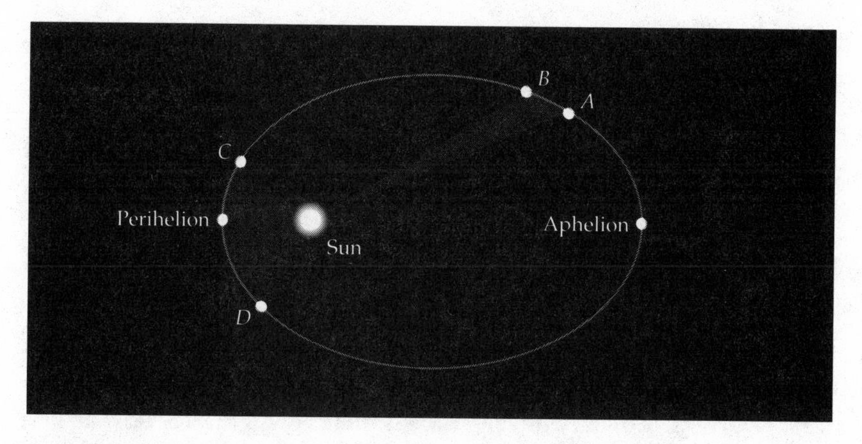

According to what came to be termed Kepler's first two laws, every planet travels around the sun along an elliptical orbit with the sun at one focus in such a way that the line joining the planet and the sun sweeps out equal areas in equal intervals of time.

At first glance, Kepler's description of his discoveries in *Astronomia nova* seems a straightforward account of a lengthy journey into unknown territory, enlivened by a highly personal, almost confessional, style punctuated by rhetorical flourishes. He appeared to spare no detail, recounting every hypothesis made and rejected, every false trail followed and then abandoned, every computational failure and triumph, until his final convergence on the truth.

He complained to his readers. "If this cumbersome mode of working displeases you, you may rightly pity me, who had to apply it at least seventy times with great loss of time." He expressed continual skepticism. "Whoever would think it possible? This hypothesis, which so closely agrees with the observed oppositions, is nevertheless false." He passed judgment. "The sun will melt all this Ptolemaic apparatus like butter, and the followers of Ptolemy will disperse partly into the camp of Copernicus, partly into that of Brahe." He scolded himself. "I thought and searched, until I went nearly mad, for a reason why the planet preferred an elliptical orbit. . . . Ah, what a foolish bird I have been."

Many historians have taken Kepler at his word and accepted *Astronomia nova* as a remarkably candid account of his travails. But that view is starting to change. In 1973, Owen Gingerich wrote, "Most commentators have assumed because of Kepler's sequential and at times autobiographical style that Kepler spared no detail in the chronicle of his researches. Examination of the manuscript material shows on the contrary that the book evolved through several stages and represents a much more coherent plan of organization than a mere serial recital of investigations would."

In other words, Kepler's goal was to persuade his readers—a highly sophisticated, technically proficient, knowledgeable group of practitioners—that there was no other course possible than to accept his radical theories. He chose a style and approach designed to demonstrate clearly that every conceivable alternative, no matter how attractive initially, ultimately fails the observational test.

As Bruce Stephenson shows in *Kepler's Physical Astronomy*, a recent detailed analysis of the physics and mathematics contained in *Astronomia nova*, Kepler organized his book to meet every objection that his readers might have. He planted suspicions, asked provocative questions, and then provided ingenious solutions, all in his inimitable, idiosyncratic style. Kepler expected resistance, and he wanted to persuade skeptics that his astronomy was not merely an acceptable way of computing planetary positions but represented an accurate model of the physical solar system.

To modern minds, his modus operandi appears exceedingly clumsy and complicated, and he spends an inordinate amount of time on issues no longer relevant. But those were the issues that were important in the early seventeenth century, and Kepler had to use mathematics and notation familiar to his readers. He shuffled and rewrote chapters in such a way that a select audience of mathematicians and astronomers might be persuaded that practically all the planetary theory they knew was false, that their methods of calculation were outdated, and that his new theory was the correct alternative. One by one, he put forward detailed arguments favoring one or another mathematical apparatus, then demolished the structure by demonstrating how it failed to meet the test of Tycho's observations. In every case, Kepler insisted that his own theory supplied a simpler, more elegant answer to such questions as why Earth's motion isn't uniform and why its distance from the sun varies.

In the end, Kepler's exhaustive and exhausting calculations, recounted in such painstaking detail, demonstrated not only the ways in which traditional planetary theory failed to embody Tycho's observational data but also why that theory was nonetheless remarkably successful. In particular, the difference between an elliptical and a circular orbit was small enough (especially in the case of Earth) that a displaced, or eccentric, circle was nearly equivalent to an ellipse of low eccentricity. Of all the known planets, Mars had been the most perplexing, because its orbit is more elongated, or eccentric, than any other. It provided the strongest evidence of deviations from a circular orbit, and it was on the basis of these small but significant deviations that Kepler built his argument that an ellipse rather than an eccentric circle is the fundamental shape of a planet's orbit. For all the other planets, numerical estimates based on Copernican and Ptolemaic theory approximated the results of Kepler's area law so closely that there was little to differentiate Kepler's approach from traditional theories.

Of course, at the core of Kepler's argument was his conviction that the motions of the solar system have a physical cause. As a physical process, motion in an eccentric circle, though appealing as a mathematical model, was really quite complicated—so complicated as to raise doubts of its reality. As Stephenson notes, Kepler "performed so radical a reassessment by interpreting astronomy, for the first time, as a physical science. . . . No longer did it suffice that a theory was mathematically plausible. Mathematical elegance appealed to the astronomer, but real bodies were moved by physical forces acting on other real bodies."

When they finally appeared in print, Kepler's 70 chapters of new astronomy had little immediate impact. Intimidated more than impressed, few readers could navigate the book's mathematical shoals or accept its demotion of the sacred circle. No one seemed to understand what Kepler was up to; Galileo, though he received a copy, remained silent. In the eyes of most scholars of the time, Kepler's book probably didn't distinguish itself from other advanced treatises on astronomy. As a book reviewer of the period might have put it, the volume represented a mildly interesting, though perplexing and difficult, contribution from an accomplished mathematician and astronomer with a growing reputation. Indeed, *Astronomia nova* was just one of literally hundreds of scholarly books published annually throughout Europe.

Ellipses and orbits

The largest diameter across an ellipse is called the major axis; half that distance is known as the semimajor axis, *a*. The shape of an ellipse is determined by its eccentricity, *e*. If $e = 0$, its shape is a circle.

Increasing its eccentricity makes an ellipse more squashed. With the exception of Mercury and Pluto, the planets have orbits that are very nearly circles, and the sun is only a little off center.

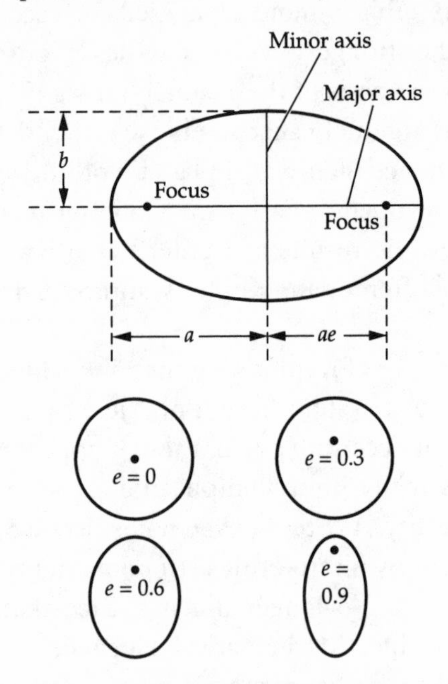

Eccentricities and Inclinations of Planetary Orbits

Planet	*Eccentricity, e*	*Inclination of Orbit to Ecliptic (degrees)*
Mercury	0.206	7.0
Venus	0.007	3.4
Earth	0.017	0.0
Mars	0.093	1.9
Jupiter	0.048	1.3
Saturn	0.056	2.5
Uranus	0.046	0.8
Neptune	0.010	1.8
Pluto	0.248	17.1

The dashed line divides the six planets known in Kepler's time from the three outer planets discovered later.

While tangling with Mars and for years afterward, Kepler produced an enormous number of books, pamphlets, tracts, and letters on all manner of topics, at one point writing the equivalent of a book a week. Despite persistent illness and family misfortunes, he somehow still had time left over not only for his interminable calculations but also for his business affairs, which included the never-ending struggle to extract a salary from an emperor who always promised more than his treasury or flunkies would allow.

The year 1611 was particularly tragic. Kepler's wife and their six-year-old son both died, and political tensions erupted into civil war in the region surrounding Prague. This local upheaval led in 1618 to the start of the Thirty Years War, a series of bitter conflicts between Catholics and Protestants that devastated central Europe. Kepler repeatedly found himself caught between opposing factions, each side vying for astrological and political advantage. Within a year he had moved to Linz to escape the turmoil, although he proved astute enough politically in his prognostications to retain the title of imperial mathematician. Meanwhile, he obtained a new job as a provincial mathematician based in Linz.

Even in the midst of domestic tragedy and political crisis, Kepler could still turn his mind to such questions as the geometry of snowflakes, the subject of a delightful little book published in 1611. One can imagine Kepler on a wintry day, crossing a stone bridge over Prague's Vltava River, stopping to contemplate the snowflakes clinging to his sleeve. Nearsighted, he would bend over to examine them closely. Why do snowflakes sometimes fall as six-pointed starlets, he asked himself, and why do all snowflakes have a sixfold symmetry? Could the answer lie in the way hard spheres pack into hexagonal arrangements? "So having examined every notion that occurred to me," Kepler wrote, "I conclude that the cause of the snowflake's six-cornered shape is the same as the cause of the ordered shapes and fixed numbers we find in plants. . . . I do not think that even in snowflakes these ordered figures occur by chance. There is a formative faculty that resides in the body of the earth and is conveyed by vapor just as spirit conveys the human soul."

Driven by a belief that all of nature is subject to decipherable mathematical laws, Kepler pursued numerical questions wherever he could find them. He was continually asking why one number rather than another had occurred.

The year 1611 also saw the publication of another of Kepler's books on optics, in which he established techniques for tracing the path of light rays through such instruments as the telescope, which was just coming into use. Kepler had received news of Galileo's telescopic observations of 1610, and he managed to borrow a telescope that Galileo had made and presented to a Bavarian noble. For about a month, Kepler was able to watch the four moons of Jupiter, and he wrote a brief pamphlet, *Observation-Report on Jupiter's Four Wandering Satellites,* in which he provided the first independent, written confirmation of Galileo's discovery. Soon after, Kepler himself invented a different type of telescope, which employed two convex lenses instead of one concave and one convex lens. It provided an inverted rather than an upright image and a greatly enlarged field of view.

As if that weren't enough, Kepler was deeply involved in promoting a wondrous new method of computation whereby simple addition and subtraction replaced the tedious steps involved in multiplying and dividing. In particular, the compilation of astronomical tables for both astrological and navigational purposes required so many arithmetical computations that logarithms, invented by Scottish landowner John Napier, created a sensation throughout Europe even before the first logarithmic tables appeared in 1614. With the enthusiastic support of such practitioners as Kepler, it didn't take long for logarithms to establish themselves as the great astronomical time-saver.

Indeed, computation was never far from Kepler's mind. In the winter of 1617, preoccupied with defending his mother, who had been accused of witchcraft, he stopped briefly in the German town of Tübingen, where he met a young professor named Wilhelm Schickard. Although Kepler was 24 years older, the two men had much in common. Both came from the same province in Germany, had attended the same university at Tübingen, and had similar interests in mathematics and astronomy. Moreover, Schickard was a skilled artisan capable of drawing fine maps and constructing high-quality instruments. The two became friends and in letters and conversation often discussed the latest developments in mathematics and natural philosophy.

Those discussions led Schickard to invent what may very well have been the first mechanical calculator, or "calculating clock" as he called it. Completed in 1623, this geared mechanism, which resembles an

MIRIFICI

Logarithmorum
Canonis deſcriptio,

Ejuſque uſus, in utraque
Trigonometria; ut etiam in
omni Logiſtica Mathematica,
Ampliſsimi, Facillimi, &
expeditiſsimi explicatio.

Authore ac Inventore,
IOANNE NEPERO,
Barone Merchiſtonii,
&c. Scoto.

EDINBURGI,
Ex officinâ ANDREÆ HART
Bibliopolæ, CIↃ. DC. XIV.

The title page of John Napier's tables of logarithms. Published in 1614, these tables were recognized immediately as a significant new computational tool of great value to astronomers. (Library of Congress.)

old-fashioned mechanical cash register, added and subtracted numbers with remarkable facility. It even included a bell to notify the user when a computation exceeded the calculator's capacity.

In a letter dated September 20, 1623, Schickard wrote to Kepler announcing his invention: "What you have done [by calculation], I

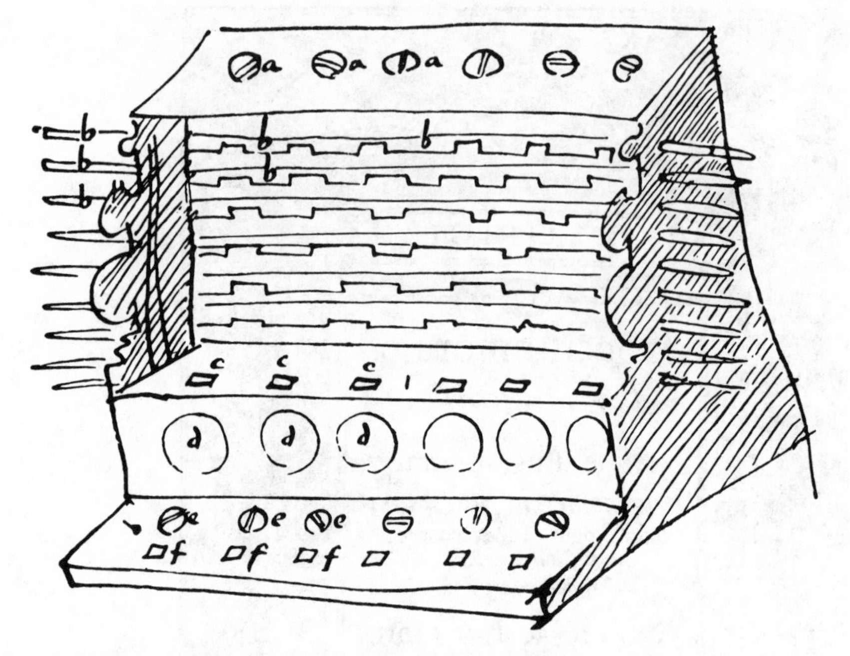

Wilhelm Schickard's sketch of his "calculating clock," which combines an adding machine with a mechanical means of representing numbers as logarithms. (Smithsonian Institution.)

have just tried to do by way of mechanics. I have constructed a machine consisting of eleven complete and six incomplete sprocket wheels which can calculate. You would burst out laughing if you were present to see how it carries by itself from one column of tens to the next or borrows from them during subtraction."

Unfortunately, Kepler never received his copy of the machine. A fire at the shop where it was being constructed destroyed the half-finished mechanism, and Schickard had no time to produce a replacement. Caught up in the devastating Thirty Years War, Schickard died of bubonic plague in 1635, and practically all traces of his invention disappeared until a few decades ago.

In the meantime, Kepler had completed *Harmonices mundi* (Harmonies of the world), which contains what is now known as his third law. This numerical relation links a planet's period—the time it takes to make a complete circuit of the sun—with its distance from the sun. What Kepler found is that for every planet in the solar system, the ratio of the square of its period to the cube of its distance from the sun is

A Demonstration of Kepler's Third Law

Planet	*Period,* T *(years)*	*Semimajor axis,* a *(astronomical units)*	T^2	a^3
Mercury	0.24	0.39	0.06	0.06
Venus	0.61	0.72	0.37	0.37
Earth	1.00	1.00	1.00	1.00
Mars	1.88	1.52	3.53	3.51
Jupiter	11.86	5.20	140.7	140.6
Saturn	29.46	9.54	867.9	868.4
Uranus	84.04	19.19	7,063	7,067
Neptune	164.8	30.06	27,159	27,162
Pluto	248.6	39.53	61,802	61,770

Kepler's third law states that the squares of the periods of the planets (in years) are equal to the cubes of their average distances from the sun (semimajor axes as measured in astronomical units). The dashed line divides the six planets known in Kepler's time from the three outer planets discovered later.

the same. For example, according to Kepler's rule, a planet 9 times more distant than Earth from the sun would take 27 years to travel once around its orbit.

Kepler wrote of his discovery in these triumphant words: "What I prophesied two-and-twenty years ago, as soon as I discovered the five solids among the heavenly orbits . . . that for which I joined Tycho Brahe, for which I settled in Prague, for which I devoted the best part of my life to astronomical contemplations, at length I have brought to light, and recognized its truth beyond my most sanguine expectations. . . . The die is cast, and I write this book. Whether it will be read by my contemporaries or by posterity is not important. If God himself has waited six thousand years for someone to contemplate his works, my book can wait for a hundred."

The discovery of the third law was a striking instance of the extraction of a general truth from a confused array of observational data. But Kepler's immediate fame ultimately rested on the *Rudolphine Tables,* finally published in 1627, just three years before he died. Containing the positions of stars, the longitudes of the world's most important cities, tables of planetary orbits, and tables of logarithms, among many other items, this volume became a standard astronomers' handbook throughout the rest of the seventeenth century. In it, Kepler showed that his revolutionary planetary theories and mathematical techniques worked better than traditional theories where it counted most: in predicting accurately the motions of the planets.

It was not until half a century later, with Isaac Newton's powerful insights into the true nature of the physical force that governs the planets, that Kepler's three laws were rescued from obscurity and enshrined among the great achievements of science. Newton showed that an elliptical orbit follows naturally from a specific, central force of gravity, providing a physical reason for why planetary orbits had to be elliptical. At the same time, Kepler's remarkable work bolstered Newton's derivations and hastened their acceptance.

Kepler occupies a unique niche in the history of astronomy and science. "It can be said of Kepler, as of few great scientists, that what he accomplished would never have been done had he himself not done it," Bruce Stephenson has argued. "The discovery from the examination of naked-eye observational reports that planets move on ellipses, and according to the area law, is so exceedingly improbable—and Kepler's manner of arriving at it was so decidedly personal—that it lies outside the course of any inevitable development."

The relentless activity, endless strife, and chaos of Kepler's own life contrast sharply with the harmony and perfection he unceasingly sought in the heavens and on Earth.

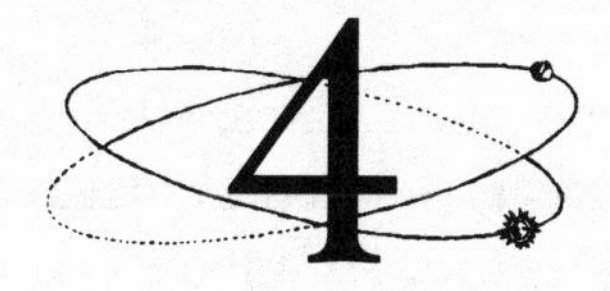

Seas of Thought

•

Where the statue stood
Of Newton, with his prism
and silent face,
The marble index of a mind forever
Voyaging through
strange seas of thought, alone.

WILLIAM WORDSWORTH (1770–1850),
The Prelude

TORN BY civil wars set against a backdrop of political upheaval, religious fanaticism, and philosophical uncertainty, England in the seventeenth century was a nation of extreme contrasts. It was a time of great wealth and dire poverty, of entrenched superstition and spirited scientific inquiry, of conservative mores and rapid social change. The growth of commerce and of the middle class, particularly in larger towns and cities, threatened traditional social hierarchies. Tradesmen amassed wealth and influence rivaling that of the nobility. Lawyers reaped the fruits of an extremely litigious age, making their mark in the courts and in politics. Doctors were learning the fundamentals of human physiology but could do little to stem the tide of devastating plagues and epidemics that swept back and forth across

Europe. Although most citizens rarely ventured beyond the fields surrounding the villages that dotted the countryside, a significant number roamed the kingdom. Others looked to the lands across the sea, founding and populating settlements in the Americas and elsewhere.

In those restless, unsettled times, even the heavens seemed in disarray. The provocative, revolutionary ideas of Copernicus, Galileo, and Kepler had set the stage for a complete reevaluation of ancient and medieval models of the universe. But no clearly superior alternative had yet emerged. With the world opened up to an unprecedented degree of scrutiny by the invention a few centuries earlier of the compass, and later of the telescope and microscope, scholars at established universities argued the merits of competing world systems. Professionals and dilettantes alike gathered in coffeehouses, clubs, and other fashionable establishments to debate the latest scientific and philosophical curiosities emerging from such investigations.

So it isn't at all surprising that in January of 1684, three erudite London professionals were engaged in a spirited discussion of why planets move in elliptical orbits. A few decades earlier, René Descartes had proposed that all of space was filled with minute, invisible particles that collectively moved in huge whirlpools, or vortices. According to Descartes's system, the sun lay at the center of one of these raging celestial whirlwinds, and its swarm of revolving particles dragged the planets along. A smaller vortex kept the moon in motion around Earth.

To Christopher Wren, Robert Hooke, and Edmond Halley, however, the whirlpool theory of Descartes was far from satisfactory. Although it provided a vivid, readily imaginable picture of the workings of the heavens, it gave no explanation of why the planets moved in ellipses with the sun at one focus and why planets nearer the sun had shorter years than more remote planets. Wren, Hooke, and Halley all preferred to think of the planets as being forcibly attracted by the sun. But what was the nature of that force?

Of the three men, Wren, at 52, was the senior. Aristocratic and politically adroit, he had been a professor of astronomy at Oxford and made substantial contributions to mathematics, physics, and physiology. But the rebuilding of London after the great fire of 1666 had given him a chance to demonstrate his brilliance as an architect. Although architecture became his chief occupation, Wren maintained a keen

Left: Isaac Newton (1642–1727). *Right:* Edmond Halley (1656–1742). (Smithsonian Institution.)

interest in scientific and philosophical matters. He had been one of the founders of the Royal Society for the Improvement of Natural Knowledge, which began in 1660 as the Invisible College for the Promoting of Physico-Mathematical Experimental Learning. The society's members included scientists, doctors, noblemen, lawyers, civil servants, literary men, and even a few tradesmen. Its headquarters in London served as a convenient meeting place for the regular exchange of news and views concerning the latest scientific novelties.

Robert Hooke, then 49, was a puny man who depended almost entirely on the Royal Society for his income. Inventive, frenetic, and querulous, he had early in life worked as an assistant to Robert Boyle, whose 1661 book *Chymista scepticus* (The skeptical chemist) had introduced the concepts of element, alkali, and acid into discussions of the chemical composition of matter. Hooke was at various times a fellow and an employee of the Royal Society, serving as its curator. His contract specified that every week (except during summer vacation), he had to perform three or four experiments demonstrating the application of natural laws to various phenomena. It was his duty to keep track of what was happening in natural science, noting new discoveries, devising experiments, and providing his own speculations. Indeed,

Hooke belonged to an age when scientific discovery seemed an everyday occurrence, and individuals from many different walks of life could make significant contributions. Ingenious, eager, often working well past midnight, he flitted from topic to topic, one week studying the way materials stretch, the next week using an improved reflecting telescope to search the heavens. But he rarely had time to dwell on his discoveries or to develop his ideas in detail, continually pushed as he was to a new topic in preparation for the next Royal Society gathering.

Although only 28, Edmond Halley had already made his mark as an observational and theoretical astronomer. In 1676, while still a student at Queen's College, Oxford, he had obtained a year's leave of absence to visit the remote Atlantic island of St. Helena in order to map the stars visible from the Southern Hemisphere. That same year, he had published his first paper, in the *Philosophical Transactions of the Royal Society,* concerning an improved method for computing the values of key characteristics of the orbits of Jupiter and Saturn. Active and well-traveled, Halley already exerted considerable influence in scientific circles. Like Hooke's, his modest income came from the Royal Society, where in the humble position of clerk his primary duty was to record the society's activities.

The trio focused its discussion on the question of what path a body attracted by the sun would follow if the force of attraction diminished as the square of the distance between the sun and the body. Investigators in the previous century had already identified such a relationship in the way light weakens as the distance from a source increases, becoming, for example, one-quarter as strong when the distance doubles. It was not unreasonable to posit a similar relationship for the attraction between the sun and a planet. The problem was to prove that a force varying in strength as the inverse square of the distance would require a planet to travel in an ellipse and thereby to establish that the force behaves in a manner consistent with Kepler's laws.

Hooke believed he had the answer but failed to furnish a satisfactory mathematical proof to support it, even when spurred on by Wren's offer of a prize, in the form of a book worth 40 shillings, to anyone who could come up with the answer within a few months. Lacking the mathematical techniques necessary to solve exactly the equations of motion derived from an inverse-square force law, Hooke could find only approximate, numerical answers that produced orbits resembling ellipses. But he couldn't prove that his curves truly met the mathematical definition of an ellipse, though he insisted otherwise.

A visit to Cambridge in August gave Halley the opportunity to pursue this unresolved matter further. He consulted Isaac Newton, who had already earned a reputation for his quick mind and mathematical ingenuity as a professor at Trinity College. Secretive, irritable, sensitive to criticism, and curiously humorless, Newton had virtually isolated himself from the world that lay beyond his college quarters. Veering between exhilaration and exhaustion, he studied and pondered whatever topics happened to catch his attention, pursuing them with such concentrated intensity that he rarely had time for recreation and often sacrificed sleep, personal appearance, and meals to keep after his prey.

Humphrey Newton, who served as his assistant and secretary during a portion of Newton's academic career, later furnished these personal details: "His carriage then was very meek, sedate, and humble. . . . I cannot say I ever saw him laugh but once. . . . I never knew him to take any recreation or pastime, either in riding out to take the air, walking, bowling, or any other exercise whatever, thinking all hours lost that was not spent in his studies. . . . He very rarely went to dine in the Hall except on some public days, and then, if he had not been minded, would go very carelessly, with shoes down at heels, stockings untied, surplice on, and his head scarcely combed."

Newton's academic duties required him to give lectures on arithmetic, geography, optics, and other topics. These lectures were given at least once a week at a prearranged time during the seven months of the Cambridge academic year and usually lasted about half an hour. Because Newton's lectures quickly gained the reputation of incomprehensibility, there was often no audience. In addition, like all the other professors, he had to make himself available for a two-hour period twice a week during term to answer questions and help with problems.

Despite Newton's isolation and peculiar temperament, news of his extraordinary mathematical explorations and extensive experimental and theoretical studies in mechanics and optics slowly leaked out. He corresponded with members of the Royal Society and other individuals, casually dropping tantalizing hints of his far-ranging discoveries. Colleagues at Trinity privately passed on what they could glean of his innovative work in mathematics. In 1671, Newton himself sent the

Royal Society a model of the reflecting telescope he had invented. It created a sensation and ensured his election as a fellow of the society. But his first paper, published in the *Philosophical Transactions* a year later and setting forth a new theory of light and color, met with considerable skepticism and opposition. The criticism discouraged a wounded Newton from making additional forays into the public arena for more than a decade.

Soon after, Newton turned his attention to alchemy and theology, expending the same prodigious effort on these pursuits that he had on mechanics, mathematics, and optics. In a famous 1679 letter written in reply to a note from Hooke, he even admitted, perhaps somewhat ingenuously, that he no longer had any sustained interest in mathe-

A replica of Newton's reflecting telescope. (Smithsonian Institution.)

matical and scientific investigations. He regarded the universe as a secret to be solved by applying pure thought to certain clues that had been left behind by the Creator, clues that resided partly in the natural world and partly in certain mystic traditions and papers handed down since Babylonian times.

Guarded but hospitable, Newton welcomed Halley, 13 years his junior, to his quarters, perhaps offering brief glimpses of his garden, laboratory, and telescopes to his curious visitor. It's highly unlikely, however, that Newton talked at all about the alchemical and theological matters that preoccupied him at the time. But when the conversation turned to celestial mechanics, Halley took the opportunity to raise the issue that still perplexed him. What is the shape of the curve traced by a planet attracted by a force obeying an inverse-square law?

Newton replied immediately that the curve had to be an ellipse. Amazed by Newton's prompt, unequivocal answer, Halley asked for more details. Newton mentioned that he had already solved the problem, but with his customary reticence when it came to making his calculations and ideas public, he merely promised to send his demonstration to Halley at a later time. In this case, Newton's restraint was justified. Repeating the derivation, he saw that it didn't quite work. He found the source of the error in a hastily drawn diagram and proceeded to compose a fresh proof.

Newton's reply reached Halley in November of 1684 in the form of a nine-page paper called "De motu corporum in gyrum" (On the motion of bodies in an orbit). Much more than simply an answer to the question of the shape of orbits, the paper greatly impressed and delighted Halley. Although he probably didn't grasp its full significance at once, Halley could readily see that the treatise was out of the ordinary, demonstrating not just one but all three of Kepler's laws in a striking synthesis of earthly and celestial motions. Within days, he was off to Cambridge to confer with Newton and urge him to write a version for presentation to the Royal Society.

Halley expected to receive the reworked treatise, with its revolutionary implications, in a matter of months. But Newton insisted on taking as long as he felt necessary to follow his reasoning to its necessary conclusions. The effort involved in writing the paper rekindled Newton's interest in mechanics, and he eagerly plunged into a detailed, comprehensive treatment of dynamical astronomy. For nearly two years, the mathematical intricacies of mechanics commanded his attention to the exclusion of all else. Often lacking sleep and food, he single-mindedly threaded his way through the mathematical maze posed by the motions of the planets. Step by step, Newton's brief tract expanded into the majestic, immortal *Philosophiae naturalis principia mathematica* (The mathematical principles of natural philosophy).

The Royal Society officially received Newton's manuscript of what would become the first book of the *Principia* on April 28, 1686, and ordered Halley to arrange for printing the document. The society itself, dependent on fees from fellows and bequests from wealthy patrons, was nearly bankrupt after funding an elaborate, costly edition of *Historia piscium* (The history of fishes), which had sold rather poorly. Newton refused to help pay for the publication of his own work, so the burden of financing the venture fell to Halley. Indeed, supervising the publication of the *Principia* became Halley's full-time occupation, as he shuttled continually between Newton and the printer to keep the massive work on schedule.

Newton completed the remaining two books of the *Principia* in the autumn of 1686, but a dispute between Newton and Hooke nearly derailed the venture. Hooke wanted some acknowledgment that it was he who had originated the notion that planetary motion could be explained on the basis of an inverse-square gravitational force, which he had mentioned in his 1679 letter to Newton. Although Newton hadn't responded to the idea at the time, it's probable that the suggestion had triggered the calculations that Newton later consulted when Halley posed his question at their meeting. Newton could not accept Hooke's claim and, displaying his sensitivity to criticism and the vindictive, manipulative side of his character, threatened to withhold the third part of the *Principia.* A recalcitrant Newton complained to Halley, "The third I now designe to supress. Philosophy is such an impertinently litigious Lady that a man has as good be engaged in Law suits as have to do with her. I found it formerly so & now I no sooner come nearer her again but she gives me warning."

Halley eventually coaxed and flattered Newton into completing his work, and on July 5, 1687, was no doubt greatly relieved to announce in a letter to Newton that the printing of the book's 511 pages was finally complete.

Newton never granted Hooke the credit he sought. He had not forgotten Hooke's condescending treatment of his hard-won theory of light when it was presented to the Royal Society in 1672. Hooke's lengthy but hasty critique rankled for years afterward. Although the two men continued to correspond sporadically, Newton maintained a

PHILOSOPHIÆ

NATURALIS

PRINCIPIA

MATHEMATICA.

Autore *JS.* NEWTON, *Trin. Coll. Cantab. Soc.* Matheſeos Profeſſore *Lucaſiano*, & Societatis Regalis Sodali.

IMPRIMATUR.

S. PEPYS, *Reg. Soc.* PRÆSES.

Julii 5. 1686.

LONDINI,

Juſſu *Societatis Regiæ* ac Typis *Joſephi Streater.* Proſtat apud plures Bibliopolas. *Anno* MDCLXXXVII.

The title page of the first edition of Isaac Newton's imperishable *Philosophiae naturalis principia mathematica* (Mathematical principles of natural philosophy), published in 1687. (Smithsonian Institution.)

prickly isolation, while Hooke, in his usual manner, plunged into one scientific venture after another.

In a somewhat limited sense, Hooke was perhaps right in claiming that he had considered the inverse-square law before Newton had, but he had failed to develop the idea. However, much more was at stake in the bitter dispute between Newton and Hooke than the inverse-square law. It involved competing systems of philosophy to explain how the world worked, a matter that both Newton and Hooke took very seriously. Hooke, in fact, went to uncharacteristic lengths to prepare a series of lectures delivered at various times in 1687 and 1688 as a reply to Newton and as a defense of his own grand synthesis. These "Lectures of Earthquakes" featured the image of a spinning Earth flattened at the poles, with Earth's shape tied to a complex cosmogony and general law based on changes in Earth's axis of rotation. Hooke believed that Earth's shape was a true reflection of its dynamical history.

Newton, of course, saw these lectures as a direct threat to his own system, the subject of the *Principia.* But the extravagant praise showered on the book at its publication ensured Newton's status as a mathematician, and it was Newton's highly mathematical, coherent vision of the universe that changed the course of mathematical physics and set the standard of scientific discourse for centuries to come. His brilliant scheme, disseminated over the years by eager disciples, captured the spirit of a new age in which rational discourse played a central role. Greatly overshadowed by Newton's system, Hooke's ingenious but limited scheme quickly faded from view.

In developing his unified vision of the world, Newton benefited greatly from the mathematical insights of such prominent scholars as François Viète, René Descartes, Pierre de Fermat, Blaise Pascal, and Newton's predecessor at Cambridge, Isaac Barrow. The early part of the seventeenth century had seen important developments in algebra, including the use in equations of letters and other symbols signifying quantities in place of specific numbers. This compact, suggestive notation gradually transformed the look and very nature of mathematics. Using algebraic formulas, mathematicians could develop general rules for computing mathematical quantities such as the logarithms of numbers.

Mathematicians had also learned how to solve specific types of equations, working out step-by-step strategies, or algorithms, for determining the values of unknown quantities. And they had found ways of linking algebra and geometry so that, by specifying the coordinates of points on a curve, they could use algebraic equations to describe and plot those curves on paper.

There was also considerable interest at the time in the two fundamental and complementary problems of determining tangents to curves and finding the areas enclosed by curves. In Newton's day, mathematicians knew how to draw tangents to a variety of curves. They had also amassed a grab bag of techniques for evaluating the areas encompassed by curves, a process known as quadrature. Archimedes had approximated such areas numerically by dividing a given geometric figure into small pieces whose areas could be computed easily. Kepler had used a similar approach when he struggled to calculate the areas swept out by planets over specific periods of time.

Mathematicians were slowly coming to realize that drawing tangents to curves (what we now call differentiation) and determining the areas enclosed by curves (what we now call integration) are in fact related. This relationship, which is that of inverse operations, is now

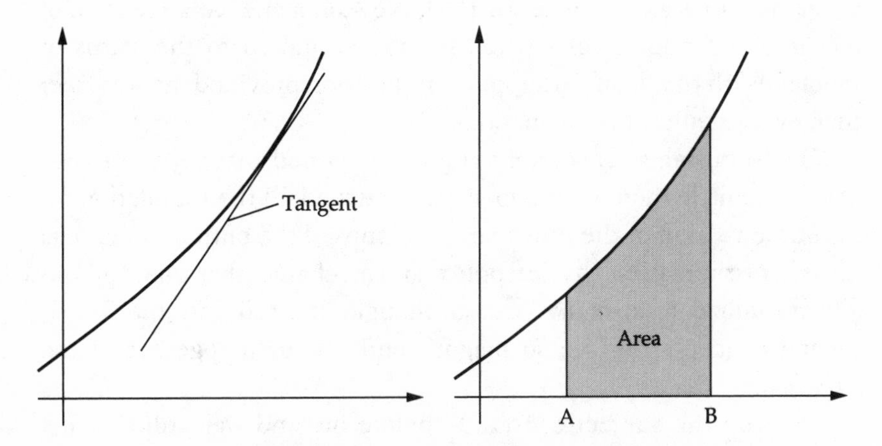

A tangent is a line touching a curve at only one point. Mathematicians often face the problem of determining which line is the tangent at any given point on a curve. Similarly, to solve certain problems, mathematicians need to determine the area between a curve and the horizontal axis, given any two points along the horizontal axis. The calculus, via the operations of differentiation and integration, provides handy techniques for finding both tangents and areas.

known as the fundamental theorem of calculus. On this key principle rest the mathematical techniques that now lie at the heart of much of science.

Books in the well-stocked libraries of Cambridge first brought these ideas to Newton's attention when he entered Trinity College as a student in 1661. With a voracious appetite for information on any topic that captured his interest, Newton would spend hours patiently working through the mathematical intricacies contained in the works of Descartes and others, nearly always without the help of a single tutor or professor. A venerable, established institution, Trinity College had at that time settled into a period of lethargy and political intrigue. Many professors, appointed for political or religious reasons, neglected their duties, and most students spent more time carousing in the ubiquitous public houses than on their studies. They led a privileged, sheltered life involving little intellectual exertion.

Strongly self-motivated and openly disdainful of his fellow students, Newton took advantage of the freedom afforded by such a loose system and plunged into whatever topics interested him. He advanced rapidly to mastery of some of the most sophisticated, up-to-date ideas in mathematics and science. In 1664, Newton achieved a measure of recognition for his efforts when he was promoted to the status of scholar. With the financial support this position provided, he was freer than ever to follow his inclinations.

Thus, during a remarkably productive period that began in 1664 and lasted little more than two years, Newton laid the foundation for his unified vision of the universe. He glimpsed the outlines of crucial ideas in mathematics, physical optics, and mechanics that would gradually crystallize, after many years of thought and toil, into the fundamental concepts that he so magnificently brought together in the *Principia*.

"I keep the subject constantly before me and wait 'till the first dawnings open slowly, by little and little, into a full and clear light," Newton once remarked. This was no drama of divine revelation, as the legends that later surrounded him would have it. Newton's "miraculous year" of 1666 provided only the first glimmerings of astonishing ideas that would gradually mature and then irrevocably change the face of mathematical physics.

DE ANALYSI

Per Æquationes Numero Terminorum

INFINITAS.

Methodum generalem, quam de Curvarum quantitate per Infinitam terminorum Seriem menſuranda, olim excogitaveram, in ſequentibus breviter explicatam potius quam accuratè demonſtratam habes.

ASI AB Curvæ alicujus AD, ſit Applicata BD perpendicularis: Et vocetur $AB = x$, $BD = y$, & ſint a, b, c, &c. Quantitates datæ, & m, n, Numeri Integri. Deinde,

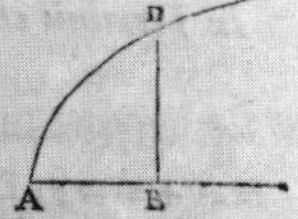

Curvarum Simplicium Quadratura.

REGULA I.

Si $ax^{\frac{m}{n}} = y$; *Erit* $\frac{an}{m+n}x^{\frac{m+n}{n}}$ = *Areæ ABD.*

Res Exemplo patebit.

1. Si x^2 $(= 1x^{\frac{2}{1}}) = y$, hoc eſt, $a = 1 = n$, & $m = 2$; Erit $\frac{1}{3}x^3$ = ABD.

A 2. Si

Written in 1669 but first printed in 1711, Newton's *De analysi* introduced many of the operations that beginning calculus students now learn. The book's opening page shows the familiar rule for integrating a power of *x*. (Library of Congress.)

A half century afterward Newton would recall: "In the same year [1666] I began to think of gravity extending to the orb of the Moon, and having found out how to estimate the force with which a globe revolving within a sphere presses the surface of the sphere, from Kepler's rule of the periodical times of the planets being in a sesquialterate proportion of their distances from the centres of their orbs, I deduced that the forces which keep the planets in their orbs must [be] reciprocally as the squares of their distances from the centers about which they revolve: and thereby compared the force requisite to keep the Moon in her orb with the force of gravity at the surface of the Earth, and found them [to] answer pretty nearly."

In other words, Newton assumed that the gravitational attraction a body exerts would depend on its mass, and a huge mass like the sun would be able to pull much harder than the significantly smaller masses of the individual planets.

Of course, Newton was not alone in considering these matters. The mechanics of uniform circular motion were well known by the 1670s, and speculations concerning the force of gravity abounded. In mathematics, tangent problems and quadratures were all the rage. Indeed, this great activity and interest meant that the world at large was now predisposed to recognize Newton's unique achievement in unifying these diverse elements into a coherent system of the world.

It all came together in the *Principia.* Newton returned to his early work, which he had refined and enlarged upon but never really completed in any satisfactory fashion during the intervening 20 years, and composed the first comprehensive treatment of forces and the resulting motions of bodies. He succeeded in fitting terrestrial and celestial mechanics into the same fundamental framework of mathematically expressed physical laws. Kepler's three laws necessarily flowed from the same principles of dynamics that Galileo had expounded for the motion of an accelerated object, whether falling to the ground or speeding over a smooth surface.

The *Principia* now stands as a prime example of the art of persuasion in a scientific text. Proceeding with characteristic caution, Newton took as his model the formal approach generally used by scholars in ancient Greece and particularly evident in Euclid's classical treatise on geometry, the *Elements.* Both works begin with definitions and axioms, then proceed step by step through a logical sequence of propositions, theorems, problems, lemmas, corollaries, explanatory notes, and examples.

But this admirably rational approach didn't guarantee a work that was easy to understand, either in Newton's era or our own. Newton was struggling with several concepts that had never before been carefully defined. He had to think through what he really meant by mass (quantity of matter), momentum (quantity of motion), inertia (passive force), and various types of active forces. He also had to assert the existence of an immovable, absolute frame of reference against which all other distances and times could be measured, carefully distinguishing this absolute space and time from the relative space and time apparent to the senses.

Newton himself frankly confessed that he had composed the *Principia* so as to test the worthiness of his readers. He wanted to ensure that they properly appreciated his painstakingly constructed logic. In much the same way that Kepler tried to prove the necessity of his conclusions, Newton hoped to forestall potential objections and criticisms. Of course, Kepler's approach, which was to throw in everything but the kitchen sink, contrasts sharply with Newton's comparatively lean, economical style.

Newton acknowledged these motives in Book III when he wrote, "But afterwards, considering that such as had not sufficiently entered into the principles could not easily discern the strength of the consequences, nor lay aside the prejudices to which they had been many years accustomed, therefore to prevent the disputes which might be raised upon such accounts, I chose to reduce the substance of this Book into the form of Propositions (in the mathematical way), which should be read by those only who had first made themselves masters of the principles established in the preceding Books." A stern taskmaster indeed!

Furthermore, the great haste with which he composed the *Principia* resulted in flaws. Derek T. Whiteside, editor of Newton's mathematical papers, has argued that "the logical structure [of the *Principia*] is slipshod, its level of verbal fluency none too high, its arguments unnecessarily diffuse and repetitive and its very content on occasion markedly irrelevant to its professed theme." Moreover, historian Richard S. Westfall has pointed out that Newton wasn't entirely immune to the temptation to adjust calculations and data to fit his preferred theories.

To modern eyes, the mathematical complexities and Newton's unfamiliar notation create a somewhat misleading impression. Mathematics in Newton's time was already remarkably sophisticated. Russian mathematician Vladimir I. Arnol'd has pointed out that Newton and his contemporaries, though lacking many of the advanced mathematical tools that mathematicians now take for granted, could solve in a few minutes numerous problems that a majority of present-day mathematicians would have to struggle with far longer to solve.

Despite the mathematical language that Newton employed, the publication of the *Principia,* fueled by Halley's promotion of an atmosphere of keen anticipation, created a tremendous stir. Whether or not the book was actually read, the philosophy it embodied had an immediate impact, and it colored philosophical discourse ever afterward. It also represented the true beginning of theoretical physics.

The book's introduction contains the famous three laws of Newton, the seeds of which had appeared in many earlier writings, especially those of Galileo Galilei, René Descartes, and Pierre Gassendi. Newton's contribution was the expression of these three laws of motion in a terse, quantitative form, gathered together to form the foundation of theoretical mechanics. As part of this formulation, he presented a novel and invaluable definition encapsulating the concept of "quantity of motion." He stated it (in Latin) as follows: "The quantity of motion is the measure of the same, arising from the velocity and quantity of matter conjointly." In modern terms, this means that every object has an associated quantity, called momentum, obtained by multiplying its mass and its velocity.

The first law, derived from the investigations of Galileo, Descartes, and others, specifies that "every body continues in its state of rest, or of uniform motion in a [straight] line, unless it is compelled to change that state by forces impressed upon it." Often called the law of inertia, this statement expresses the counterintuitive fact that, contrary to the evidence of everyday experience that all things—from a stone rolling down a hillside to a spinning top—will eventually stop moving, no force at all is required to sustain motion, and that a moving body naturally travels at a constant speed in an unchanging direction. This uniform motion persists in the absence of such forces as air resistance and friction.

Newton and his contemporaries understood the importance of the law of inertia in explaining the motion of a planet around the sun. In the simplest possible case—that of a circular orbit—the planet at each instant moves uniformly along a tangent to its orbit; simultaneously, it is drawn by the ever-present attraction of the sun. These two motions

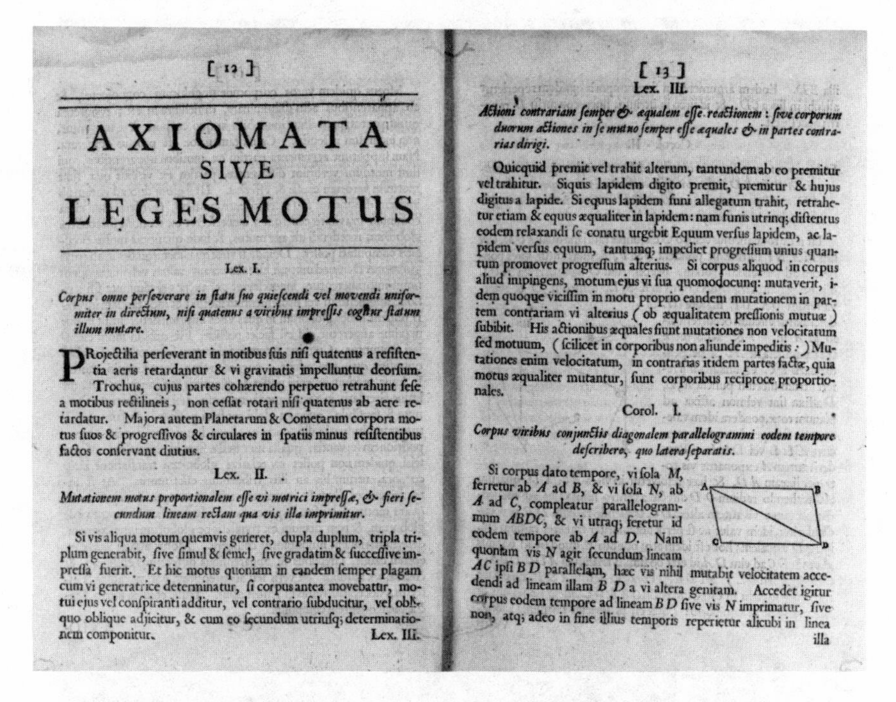

[12]

AXIOMATA SIVE LEGES MOTUS

Lex. I.

Corpus omne perseverare in statu suo quiescendi vel movendi uniformiter in directum, nisi quatenus a viribus impressis cogitur statum illum mutare.

Projectilia perseverant in motibus suis nisi quatenus a resistentia aeris retardantur & vi gravitatis impelluntur deorsum. Trochus, cujus partes cohærendo perpetuo retrahunt sese a motibus rectilineis, non cessat rotari nisi quatenus ab aere retardatur. Majora autem Planetarum & Cometarum corpora motus suos & progressivos & circulares in spatiis minus resistentibus factos conservant diutius.

Lex. II.

Mutationem motus proportionalem esse vi motrici impressæ, & fieri secundum lineam rectam qua vis illa imprimitur.

Si vis aliqua motum quemvis generet, dupla duplum, tripla triplum generabit, sive simul & semel, sive gradatim & successive impressa fuerit. Et hic motus quoniam in eandem semper plagam cum vi generatrice determinatur, si corpus antea movebatur, motui ejus vel conspiranti additur, vel contrario subducitur, vel obliquo oblique adjicitur, & cum eo secundum utriusq; determinationem componitur.

[13]

Lex. III.

Actioni contrariam semper & æqualem esse reactionem: sive corporum duorum actiones in se mutuo semper esse æquales & in partes contrarias dirigi.

Quicquid premit vel trahit alterum, tantundem ab eo premitur vel trahitur. Siquis lapidem digito premit, premitur & hujus digitus a lapide. Si equus lapidem funi allegatum trahit, retrahetur etiam & equus æqualiter in lapidem: nam funis utrinq; distentus eodem relaxandi se conatu urgebit Equum versus lapidem, ac lapidem versus equum, tantumq; impediet progressum unius quantum promovet progressum alterius. Si corpus aliquod in corpus aliud impingens, motum ejus vi sua quomodocunq: mutaverit, idem quoque vicissim in motu proprio eandem mutationem in partem contrariam vi alterius (ob æqualitatem pressionis mutuæ) subibit. His actionibus æquales fiunt mutationes non velocitatum sed motuum, (scilicet in corporibus non aliunde impeditis:) Mutationes enim velocitatum, in contrarias itidem partes factæ, quia motus æqualiter mutantur, sunt corporibus reciproce proportionales.

Corol. I.

Corpus viribus conjunctis diagonalem parallelogrammi eodem tempore describere, quo latera separatis.

Si corpus dato tempore, vi sola *M*, ferretur ab *A* ad *B*, & vi sola *N*, ab *A* ad *C*, compleatur parallelogrammum *ABDC*, & vi utraq; feretur id eodem tempore ab *A* ad *D*. Nam quoniam vis *N* agit secundum lineam *AC* ipsi *BD* parallelam, hæc vis nihil mutabit velocitatem accedendi ad lineam illam *BD* a vi altera genitam. Accedet igitur corpus eodem tempore ad lineam *BD* sive vis *N* imprimatur, sive non, atq; adeo in fine illius temporis reperietur alicubi in linea illa

Newton's three laws of motion, as stated in Latin in his *Philosophiae naturalis principia mathematica.* (Library of Congress.)

combine in such a way that the planet traces out a circular path. Planets and comets, meeting less resistance in emptier, wider spaces, preserve their motions for much longer than ordinary terrestrial objects.

The second law states that "the change of motion is proportional to the motive force impressed; and is made in the direction of the right line in which the force is impressed." In other words, Newton related force to a change in momentum (not to acceleration, as most contemporary physics textbooks state). If the mass remains constant, a change in motion is then equivalent to a change in velocity; that is, to the acceleration of a material object. Thus, all forces cause acceleration (a change in the speed or direction of an object), and doubling the force means doubling the acceleration. Interestingly, Newton's second law also describes a situation in which the mass of an object changes. This type of motion didn't become important until the advent of rockets, which propel themselves by spewing out hot gases and thus continuously change their mass.

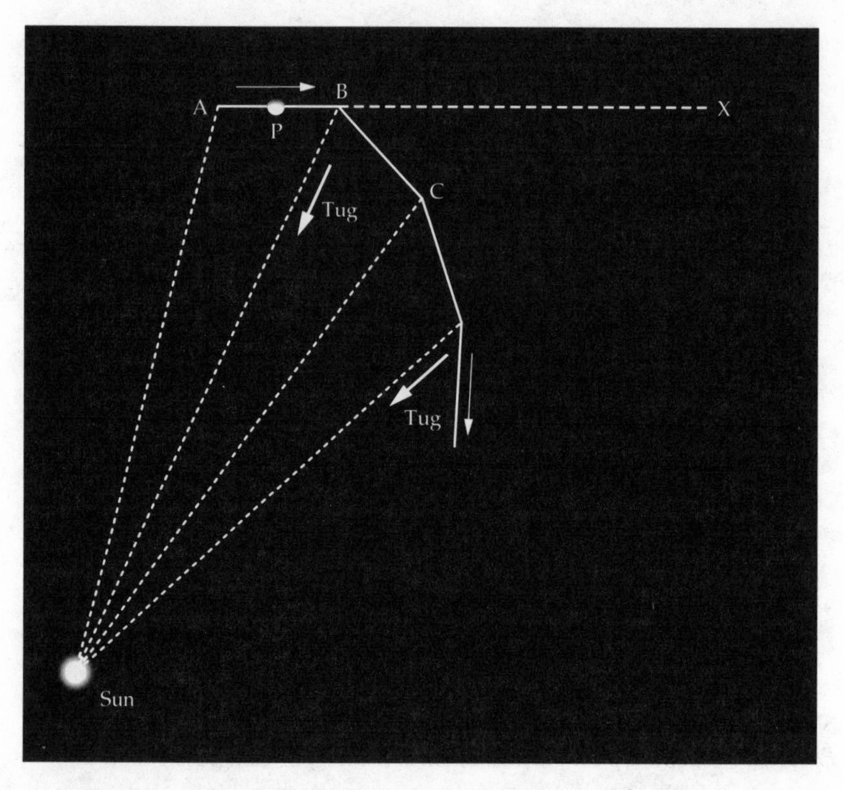

How a sequence of discrete tugs of attraction influences a planet's motion. Without a tug at B, the planet P would move off in the direction of X. When the gravitational force exerted by the sun acts continuously, the planet takes an elliptical path.

The importance of Newton's choice of momentum as a key concept in dynamics lies in the notion that momentum is one of two quantities that, taken together, yield everything there is to know about a dynamical system at a given instant. The second quantity is simply position, which determines the force's strength and direction. Newton's insight concerning the pairing of momentum and position was placed on firmer ground more than a century and a half later by the mathematicians William Rowan Hamilton and Karl Gustav Jacob Jacobi. It is a duality that lies at the heart of modern dynamical theory.

The third law says that "to every action there is always opposed an equal rection: or, the mutual actions of two bodies upon each other are always equal, and directed to contrary parts." Often misunderstood, this law probably arose from Newton's studies of colliding bodies. Imagine two identical balls suspended side by side so that they touch each other. A remarkable exchange is produced if one ball is pulled aside and then released so that it strikes the other, stationary ball. The impact sends the stationary ball into motion, and the initially moving ball comes to a standstill. Out of such experiments, Newton drew the generalization that came to be known as the law of conservation of momentum. In other words, in the absence of an external force acting on a given system, the total momentum of the system remains constant. The third law is simply an alternative way of stating this fundamental principle.

Newton's axiomatic framework allowed him to pursue a strategy in which he could construct a simplified, idealized mathematical model of the physical system he wanted to probe—in this case, the solar system. Using mathematics, Newton could work out the consequences of certain actions and compare them with measurements and empirical observations. That comparison, in turn, would suggest ways in which the model could be adjusted and refined to achieve even greater realism. In essence, this strategy of maintaining a tight interplay between mathematical analysis and physical experience afforded a marvelously productive way of using mathematics to explain the workings of nature. Revolutionary in Newton's time, this kind of approach is taken for granted in modern research.

Newton's theory of universal gravitation formed the core of Book III of the *Principia*, which appeared at long last after two books devoted to abstract arguments that set the mathematical stage. By the end of this concluding book, Newton could contend that gravity acts on all bodies in the universe. In other words, every particle of matter in the universe attracts every other particle with a force that depends simply but precisely on the masses of the particles and the distances between them. Moreover, that universal relationship could be condensed into a brief algebraic formula expressing the fact that the gravitational force acting between two bodies is proportional to the product of the individual masses divided by the square of the distance between them.

But Newton didn't really answer the question of what makes planets move. He freely admitted that his law of gravitation specifies that the sun acts on the planets and produces orbits of the type mandated by Kepler but says nothing about exactly *how* the sun acts on the planets. Gravity itself was a mystery, even though, in Newton's mind, it was the key to all celestial motion.

The fundamental premise underlying Newton's dynamics was the notion that the positions and momenta of all particles in the world at a particular instant in time determine their future course, just as those positions and momenta are a product of their past history. More than two centuries later, Henri Poincaré would encapsulate Newton's approach in these words: "For Newton, a physical law was a relation between the present state of the world and its condition immediately after, or, in other words, physical laws are differential equations."

Because Newton cast his laws in the form of mathematical equations that implicitly included time, he could try to solve the resulting differential equations (which supply the instantaneous relationship between position, velocity, and acceleration) to deduce position from information about the forces involved. For instance, we can think of the gravitational attraction near Earth's surface as practically constant because the distance of a falling apple from Earth's center changes so little over its course. The apple's position and velocity both change, and the longer it falls, the faster it goes (ignoring the effects of air resistance). In Newton's formulation, this acceleration is constant because the force itself can be described as constant, while the speed increases steadily and the distance covered increases as the square of the time elapsed. Indeed, several decades before Newton first considered the problem, Galileo's careful experiments demonstrated just such a relationship.

By combining his laws of motion with his theory of gravitation, Newton could show that two gravitationally interacting bodies would follow trajectories in the shape of circles, ellipses, parabolas, and hyperbolas. All these curves can be obtained by slicing through a cone in different ways, thereby unveiling a beautiful harmony between dynamics and geometry.

In the case of elliptical orbits, Newton's mechanics tied the two geometrical parameters that define an ellipse—its semimajor axis (which establishes its size) and its eccentricity (which establishes its shape)—to the two dynamical parameters that define a planet's orbital motion: its energy and its angular momentum, or rotational motion. Indeed, Kepler's laws represent an application of the general prin-

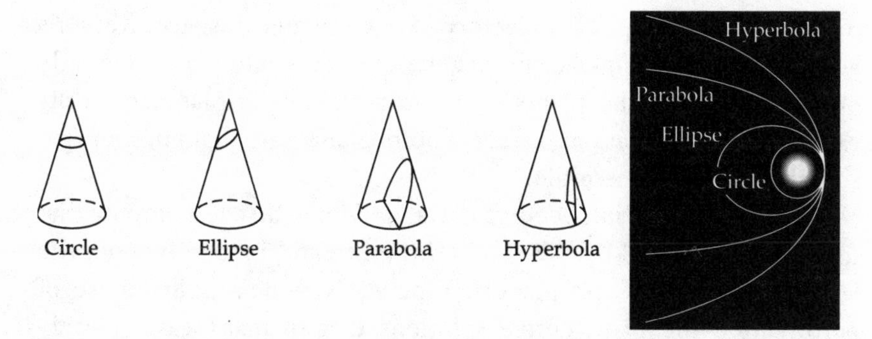

A conic section is any one of a family of curves obtained by slicing a cone with a plane, as shown. The resulting curves—circles, ellipses, parabolas, and hyperbolas—also represent the types of orbits possible when one body interacts gravitationally with another.

ciples of the conservation of angular momentum and the conservation of energy to the specific case of celestial phenomena. Of course, Newton didn't express his ideas in exactly these terms. It was only many years later that scientists developed useful definitions of energy and angular momentum, which permitted them to use those concepts in tracing celestial motions.

In general, the formulas obtained by solving the appropriate differential equations are all one needs to make predictions or deduce past history. In principle, if we knew at some fixed instant the position and velocity of every particle of matter in the solar system, we would then be able to determine all subsequent motions of those particles. Halley's use of Newton's mechanics to compute the orbit of a comet (the one named after him) was one of the first, striking applications of these new tools to the solar system.

Nonetheless, Newton's own exploratory calculations, and certainly the experiences of many later investigators, demonstrated quite unequivocally the extreme difficulty of solving the differential equations describing even so few as three bodies. To circumvent these problems, Newton inaugurated the theory of perturbations as a means of calculating complicated effects by approximation. It was clear that if Newton's gravitational law was indeed universal, then every planet in the solar system must pull on every other planet, tugging each one away from an elliptical orbit and adding a distinctive pattern of ripples to the basically elliptical trajectory of each planet in space. Kepler's laws worked amazingly well because the sun's influence so greatly outweighs that of the planets. Any irregularities in planetary orbits caused by the gravitational effect of other planets are slight in comparison, though still discernible.

Using perturbation theory, it was possible first to compute the main effects caused by the sun and then to add in, one by one, the lesser influences of the planets. Indeed, Newton's brilliant use of perturbation theory to derive solutions that in many cases closely matched observation led succeeding generations of investigators to believe that differential equations, nearly without exception, led to stable and regular motions. His published accomplishments in this area strongly reinforced the notion that objects in the universe follow unique, predetermined paths.

Newton's elegant mathematical clockwork was truly an engine of discovery and illumination. For example, Newton could attribute Earth's oblate shape to flattening at the poles caused by the planet's spinning motion. He also showed how the gradual shift in the timing of the equinoxes could be accounted for by the pull of the moon and sun on Earth's equatorial bulge, which causes the planet to precess like a spinning top.

In 1696, his last year at Cambridge, Newton agreed to produce a second edition of the *Principia,* which eventually appeared in 1713 with substantial revisions. In the meantime, perhaps aware that his mind no longer flashed as brilliantly as it once had with mathematical and philosophical invention, Newton gave up his academic pursuits in favor of a different sort of existence at the Royal Mint. There he plunged with considerable dedication into the din, heat, and long hours involved in supervising the minting of new currency. In later years, surrounded by disciples and surrogates charged with promoting his ideas and carrying forward his work, Newton became a figure of authority and adulation, completing the transformation from reclusive inventor to respected public man.

However, these years could hardly be called a time of complete calm, for Newton wasn't slow to marshal his forces and orchestrate efforts to further his own ideas and ruthlessly squelch those of his critics and competitors. He continued to work on various papers, reviving, revising, and refining insights gleaned during his days of

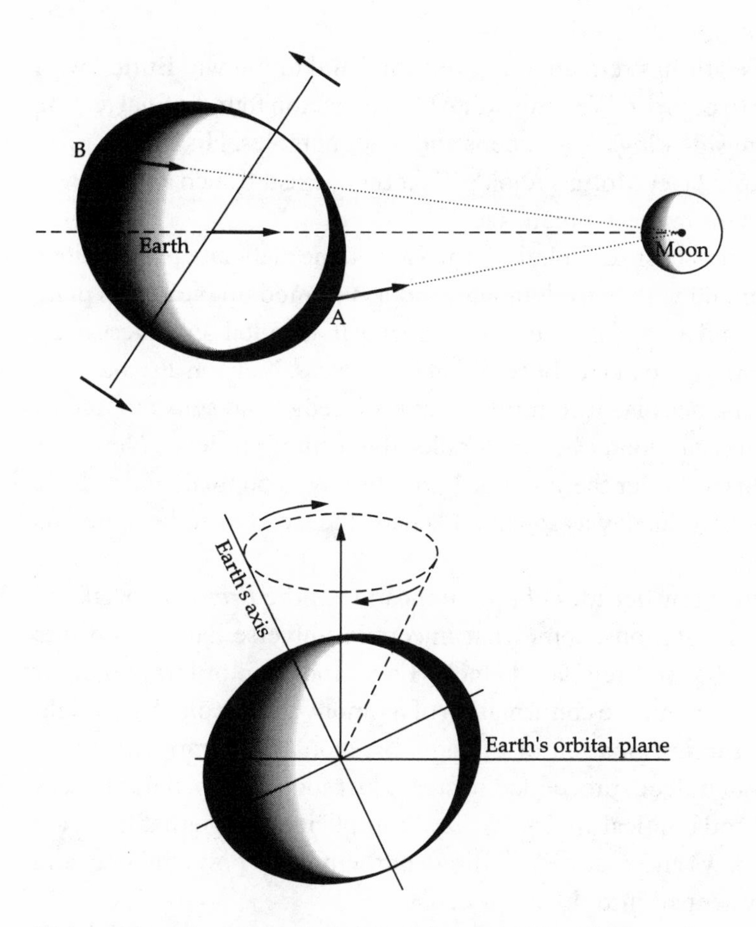

To provide a physical explanation of the precession of the equinoxes, Newton considered the moon's effect on an oblate Earth. The gravitational pull of the moon on the flattened Earth acts more strongly on the equatorial bulge on the closer side than it does on the farther side. The moon therefore tries to "straighten up" Earth's axis, which is inclined to the moon's orbit, as indicated by the arrows at the poles. But the rotating Earth reacts like a spinning top and begins to precess, and its axis of rotation slowly traces out a circle in the sky.

private scholarship many years earlier. He also enjoyed his celebrity and the power he wielded as president of the Royal Society.

Newton's death on March 20, 1727, at the age of 82, triggered an enormous response. Innumerable poems, statues, medals, and books celebrated his immense achievements, even though by far the largest part of his writings remained unpublished and unknown. Buried with appropriate pomp in Westminster Abbey, Newton found a final resting place alongside kings and queens and other notables. His tomb carries this epitaph: "Let Mortals rejoice That there existed such and so great an Ornament to the Human Race."

The French writer Voltaire, who viewed the elaborate preparations for the funeral with considerable wonder, summed up Newton's paradoxical popular appeal. "Very few persons in London read Descartes, whose works have in fact become totally useless. Newton also has very few readers, because it requires great knowledge and sense to understand him. Everybody however talks about him." Indeed, Newton's revolutionary "order theory" was undoubtedly as popular, and as little understood, in its day as so-called "chaos theory" was to become 300 years later.

Within a few decades of the publication of the *Principia*, what had seemed a mysterious, somewhat uncertain universe had settled into the guise of a well-regulated clock. The imposition of law and order on the cosmos and the confirmation of harmony that Kepler had sought captured the imagination of the age. Newton's contemporaries now sought and, indeed, proceeded to create just such a rational framework for social and political affairs. Newton's implicit message that there are universal laws and that we can discover them was a powerful one, and it quickly seeped into the culture.

In one sense, however, Newton may have been too successful. His examples of the regularity that arises from the application of simple, basic principles were extremely compelling. Consequently, for centuries afterward, few questioned the idea that, given sufficient data about initial conditions, it is possible in theory to make useful predictions. Three hundred years after the publication of the *Principia*, Hermann Bondi reflected on the blinders that Newton's achievements placed on his successors. "His solution of the problem of motion in the solar system was so complete, so total, so precise, so stunning, that it was taken for generations as the model of what any decent theory should be like, not just in physics, but in all fields of human endeavour. It took a long time before one began to understand—and the understanding is not yet universal—that his genius *selected* an area where such perfection of solution was possible."

But there were always clues that this predictability was perhaps illusory. Newton could not tame the moon, which experienced the significant gravitational attractions not only of Earth (its most important influence) but also of the sun (the next most important). Even when the much tinier effects of the other planets were ignored, it didn't appear possible to express as a simple algebraic formula this comparatively simple case of just three gravitationally interacting bodies. Similarly, Newton's perturbation methods managed to solve some, but by no means all, of the problems of the moon's surprisingly complicated motion, with its fanciful excursions from the norm. There was surely even more to mathematical mechanics than met the Newtonian eye.

No one seriously doubts the precision or the solidity of Newton's mathematical formulations of his laws. But these two-faced equations have hidden within them not only the rare instances of order that Newton so brilliantly illuminated but also the darkness of ubiquitous chaos. It took physicists and mathematicians centuries to begin to appreciate how uncommon Newton's kinds of solutions really are. Within the mathematics that encapsulates the breathtakingly simple and elegant principles of Newtonian mechanics, uncertainty lurks. In Bondi's words, "The image of the precision of Newton's solar system is so deeply imprinted on our mind that the awakening from this picture has been slow. That in itself is no mean tribute to the genius of Newton."

Clockwork Planets

•

> Then I felt like some watcher of the skies
> When a new planet swims into his ken
>
> JOHN KEATS (1795–1821),
> *Sonnet, On First Looking Into Chapman's Homer*

IN THE same year that a ragged army of American colonial rebels was savoring its decisive triumph over British troops at a place called Yorktown, a talented musician in the resort town of Bath in England was spending his nights scanning the skies. William Herschel, 42 years old, was in the midst of his second comprehensive survey of stars visible through his telescopes. By day, his four-story row house on a quiet Bath street served as a popular music school and a telescope workshop. By night, it was an observatory.

Herschel's passion for astronomy had started about eight years earlier, in 1773, after he developed an interest in mathematics. He had found that solving calculus problems in the evening was an ideal way to relax after a day filled with lessons, concerts, and other musical activities. His growing interest in mathematics led him into geometrical optics. That quickly brought him to astronomy, which soon consumed enough of his time to qualify as a serious pastime. Within a decade, Herschel became the best-equipped amateur astronomer in the world. With his considerable skill and dedication to making his instruments as perfect as possible, and with his exacting eye and great patience, his observations were highly valued among astronomers.

Herschel's ambitious goal was to determine by precise measurement whether the stars, which served as an apparently unchanging background against which astronomers could measure the shifting positions of the moon and planets, themselves changed position slightly. His novel technique for achieving this goal depended on repeatedly determining the position of a bright star relative to a nearby faint star. Herschel began by examining all the brighter stars visible through one of his home-built telescopes, which had a mirror 6.5 inches in diameter, to see if any had a faint companion.

On Tuesday, March 13, 1781, between 10 and 11 in the evening, he noticed something peculiar and made the following note: "In examining the small stars in the neighbourhood of H Geminorum I perceived one that appeared visibly larger than the rest; being struck with its uncommon appearance I compared it to H Geminorum and the small star in the quartile between Auriga and Gemini, and finding it so much larger than either of them, suspected it to be a comet."

The object appeared as a tiny but distinct bluish-green disk, about 3 arc seconds in diameter, equivalent to 1/450 of the apparent size of the full moon in the sky. Herschel knew that stars, even when viewed through the most powerful telescopes available, showed up as points rather than disks. He observed the object again four days later, after the weather had improved, using a more powerful telescope, and he noted that the disk appeared larger. This would not be the case with a stellar image. Additional observations over the next two months showed the object slowly drifting from the vicinity of the constellation Taurus into the constellation Gemini. But the disk never showed any signs of the tail that characterizes comets approaching the sun.

In the meantime, Herschel reported his discovery to Neville Maskelyne, the Astronomer Royal, who passed the news on to other astronomers. Soon a number of telescopes in various parts of Europe

William Herschel (1738–1822). (Smithsonian Institution.)

were trained on this enigmatic object. By the beginning of April, it became clear that this was no ordinary comet. It was moving too slowly, and its edges were sharp rather than fuzzy.

This newcomer to the known solar system afforded a wonderful opportunity for testing Newton's gravitational theory and laws of motion. Just 54 years after Newton's death, mathematicians and astronomers together (reflecting how intimately the two disciplines were entwined throughout this period) had built up a considerable stock of practical mathematical tools for computing orbits, particularly for predicting planetary positions and plotting comet trajectories.

By late 1781, a number of mathematicians and astronomers had independently demonstrated that the new-found object's observed positions better fitted a circular, planetlike orbit than a highly elliptical, cometlike orbit. Thus, within six months of its discovery, most astronomers were ready to accept Herschel's object as a full-fledged planetary member of the solar system. Berlin astronomer Johann Eilert Bode, convinced from the start that Herschel's object was a planet, chose the name that was eventually accepted as the planet's designation: Uranus. However, Herschel's preferred name for the planet was *Georgium Sidus* (George's Star), in line with a public relations effort by the Royal Society to flatter King George III of England into increasing his support of scientific research. This designation for the planet actually survived in British nautical almanacs published until 1847, written as "the Georgian."

The discovery of a new planet, the first since the ancients identified and began to track the five original wanderers, created a sensation. It brought Herschel fame and a new career as a professional astronomer. But he was much bothered by references to his discovery as "accidental." As he felt compelled to insist later in life, "It has generally been supposed that it was a lucky accident that brought this new star to my view; this is an evident mistake. In the regular manner I examined every star in the heavens, not only of that magnitude but many far inferior, it was that night its turn to be discovered. I had gradually perused the great Volume of the Author of Nature and was now come to the page which contained the seventh Planet. Had business prevented me that evening, I must have found it the next, and the goodness of my telescope was such that I perceived its visible planetary disc as soon as I looked at it."

Two years after the discovery of Uranus, Pierre-Simon de Laplace and Pierre François André Méchain had enough data to calculate the basic parameters of the planet's nearly circular orbit. They placed Uranus twice as far from the sun as Saturn, the planet previously believed to define the solar system's outskirts.

Today, the discovery of a new planet would prompt an immediate search of the extensive photographic records stored in the archives of observatories worldwide. This would enable its position in the sky to be traced over decades, providing the data required for a precise determination of its orbit. For Herschel's planet, astronomers had to rely on whatever records of star positions had been made by earlier observers, however incomplete or imprecise they might be. It was always possible that someone had observed the planet before but had mistaken it for a star.

The search for additional data proved rewarding for a number of astronomers. In the course of their historical detective work, they examined old star charts and catalogs for any "star" of the proper brightness that was no longer at its marked position. Calculating the motion of Uranus backward in its orbit, they could then check whether Uranus had been anywhere near where the suspect body had been spotted. Johann Bode found two such matches. The astronomer Tobias Mayer had unknowingly seen and recorded Uranus in 1756, and John Flamsteed, the first Astronomer Royal, had observed the planet in 1690. These observations alone spanned nearly a century, and they provided crucial landmarks for pinning down Uranus's orbit. The French astronomer Pierre Charles Le Monnier carefully went through his own records and was disconcerted to find that he had seen Uranus nine times before its actual discovery. Embarrassed, he published only four of the positions he had measured. We now know of at least 22 observations of this unusual "star" recorded by various astronomers during the century before Herschel's discovery.

Between old observations and the accumulation of new, more precise measurements, astronomers assumed they would be able to improve their determination of the planet's ephemerides—its predicted positions on specific dates. But they were considerably disturbed to find that Uranus wouldn't fit into their schemes. The wayward planet drifted from its predicted longitudinal position along the ecliptic by as much as 25 arc seconds in the years following its discovery.

Astronomers also had trouble resolving all the available observations into a single calculated orbit. The deviations between the calculated and observed positions seemed to get worse with the passage of time and the accumulation of new data on the planet's orbit. Even taking into account the gravitational influences of Jupiter and Saturn failed to improve the situation markedly. Alexis Bouvard, whose great talent for computation made him a worthy assistant to Laplace, complained in 1821, "The construction of the tables, then, involves this alternative: if we combine the ancient observations with the modern, the former will be sufficiently well represented, but the latter will not be so, with all the precision that their superior accuracy demands; on the other hand, if we reject the ancient observations altogether, and retain only the modern, the resulting tables will very inadequately represent the more ancient."

There seemed no way to reconcile all the data. Where was the fault? Had past observations been sloppy, or telescopes insufficiently accurate? Was there a subtle but pervasive error in the calculations or mathematical methods used? Was Newton's theory of gravitation flawed, especially when applied over such enormous distances?

Although a small number of scientists were ready to discard Newton's laws, investigators like Laplace began to speak of some "extraneous and unknown influence" that acted on Uranus. By the 1830s, the opinion that an unseen, outlying planet was disturbing Uranus was widespread, but few were in a position to pursue the question directly. Astronomers didn't know where to look, and calculations failed to provide any guidance.

By the nineteenth century, celestial mechanics had become a remote and highly specialized domain. The simplicity of Newton's laws now lay hidden within a baroque apparatus built around lengthy chains of algebraic terms far more intricate than the epicyclic machinery of Ptolemy and Copernicus. With the help of assistants, specialists spent their days painstakingly applying perturbation theory to the computation of ever finer deviations from Kepler's ideally elliptical orbits. Successive approximations yielded progressively more accurate descriptions of a planet's motion, but the horrendous calculations often took a year or longer for even a single orbit. Moreover, with so much effort expended just to get a result, there was little left over for checking the answer, even though it was easy to make mistakes that could send the computations veering far off course.

Despite its mind-numbing complexity, celestial mechanics represented a notable success story. The techniques available were sufficiently advanced by the early 1800s that, with patience, one could calculate the positions of the planets and the timing of eclipses for a century or more in advance.

Nonetheless, there were some nagging details of motion within the solar system that somehow eluded the experts. In particular, Mercury's elliptical orbit slowly rotated at a rate slightly faster than could be attributed to observational error. And, as the next chapter will explain, astronomers also had great difficulty calculating the moon's position with sufficient accuracy far into the future to meet even modest navigational needs. Finally, Uranus, continuing on its maddeningly recalcitrant course, was a particularly sharp thorn in the side of astronomy and celestial mechanics.

Although in absolute terms the difference between Uranus's actual and its expected position was quite small—a matter of an arc minute or less—celestial mechanics had advanced sufficiently far to make such deviations matter. At the same time, the deviations varied in a peculiar, suggestive manner. Until 1825, the planet was consistently ahead of its anticipated position, but as the years wore on, the discrepancy slowly decreased. In 1829 and 1830, prediction and observation agreed quite closely, and astronomers were ready to breathe a collective sigh of relief that the mystery had vanished. But the riddle reappeared later in the 1830s—with a twist. Now the planet lagged behind its predicted course.

Such puzzling antics revived suspicions that Newtonian physics itself might be at fault. One skeptic, George Biddell Airy, professor of astronomy at Cambridge, suggested that the problem lay with Newton's law of gravity. But most astronomers, impressed by the past successes of celestial mechanics, had enough confidence in Newton's

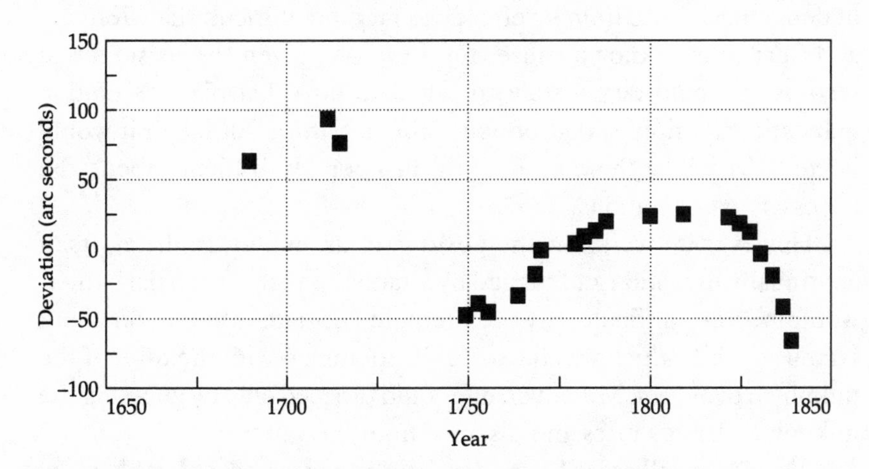

This graph shows the discrepancies (measured in arc seconds) between the observed position of Uranus and its expected position after known perturbations have been subtracted. Points for years before 1781 represent chance observations by astronomers who mistook Uranus for a star.

laws to look for an alternative solution. By 1838, they were convinced more than ever that a planet somewhere beyond Uranus was exerting just enough of a gravitational influence to cause these minor deviations. The existence of such a trans-Uranian planet became a major topic of discussion. A few mathematical astronomers even believed that they could use the intricate, cumbersome machinery of celestial mechanics to locate this mysterious body.

Airy, however, despite a strong mathematical background, rejected the notion of mathematical solutions to astronomical puzzles. He reasoned that astronomy, and indeed all science, should proceed by the collecting of observations or the taking of measurements, with mathematics used only to express the findings. Mathematics was not a tool that could tell astronomers where to point their telescopes, he insisted. Moreover, given the paucity of observations available, celestial mechanics was unequal to the task of calculating the position and precise orbit of Uranus and of deducing from that information the position of another, hidden planet. Cautious and conservative, Airy counseled waiting until Uranus had made several revolutions of the sun and its orbit was better known. Because each revolution takes 84 years, that would have entailed a lengthy delay!

Of course, Airy was right in arguing that the mathematical problem of deducing a cause from its effect was far more difficult than deriving an effect from a known cause. For example, given the existence of Uranus, it would be possible to calculate how that planet's motion gravitationally ripples that of its neighbor, Saturn. Although it would be tedious work, there is nothing insuperably difficult about the necessary computations.

However, to work out the position of an unknown planet, itself continually in motion and tugged by Uranus and other celestial bodies, would be very difficult. It would require accurate information about Uranus's orbit, which was inextricably entangled with the orbit of the unknown disturber. Moreover, one could proceed only by guessing the unknown planet's mass and distance from the sun.

That was a tall order for mathematical analysis, considerably more challenging than Kepler's untangling of the motion of Mars. Although the mathematical tools available in the nineteenth century were vastly superior to those that Kepler had at his disposal, they were still severely limited by the amount of computation required. Despite the immediate glory guaranteed anyone who could perform the calculations needed to find this elusive planet, most mathematicians wrote off the problem as insoluble or as too likely to jeopardize their careers. If they failed, they would have nothing to show for years of intense effort.

In 1841, however, this tantalizing conundrum captured the receptive mind of a gifted 22-year-old mathematics student at Cambridge. John Couch Adams was naturally drawn to astronomical puzzles. In 1835, he had experienced the thrill of seeing Halley's comet make an appearance precisely on schedule. That demonstration of the power of mathematics in astronomical prediction had a lasting effect on young Adams. Five years later on June 26, while browsing in a bookstore, he happened to pick up a copy of Airy's 1832 discussion of the Uranus dilemma. The problem made such an impression that Adams vowed to take on the task, as soon as he finished his degree, of locating by calculation the unknown planet disturbing the orbit of Uranus.

Emerging two years later as the top mathematics scholar at Cambridge, and settled into an academic career as a fellow of St. John's College, Adams pursued the Uranus calculations during summer vacations and the brief intervals between lengthy tutoring sessions and other college duties. Because he had to start somewhere, Adams initially assumed that the unknown planet lay about twice the distance of Uranus from the sun, conforming to the pattern of distances between the known planets. He proceeded by making successively finer approximations, first using circular orbits for Uranus and the unknown planet, then gradually elongating the circles into ellipses and adding other gravitational influences. His preliminary analyses strongly suggested that a planet did lie somewhere beyond Uranus. By the middle of September 1845, after successfully merging the latest observations of Uranus with data from old records, Adams had a solution that clearly pointed to a specific patch of sky where the planet ought to be found.

Adams presented his result to James Challis, director of the Cambridge Observatory, and to Airy, now Astronomer Royal. Both men had been aware of Adams's effort and had encouraged him and provided useful information. Now they seemed reluctant to take his calculations seriously enough to put aside other duties and interests to verify his prediction. Pained by this official indifference and frustrated by a succession of delays, Adams must have felt deeply discouraged. By that time, he also had a rival.

Urbain Jean Joseph Le Verrier, 34 years old and already an accomplished, internationally known mathematical astronomer in Paris, had also decided to unravel the Uranus mystery. Le Verrier had started his career in chemistry, but his superior mathematical skills turned him to astronomy. For his first major challenge, he took on the problem of the innermost planet, Mercury, which was always consistently ahead in its orbit of where calculations predicted it should be. With the tenacity for which he was well known, Le Verrier worked on this problem for three years and managed to account for most of the deviation in terms of the perturbing gravitational influence of other planets. But that still left a small discrepancy that even Le Verrier's prodigious efforts

Urbain Jean Joseph Le Verrier (1811–1877). (Yerkes Observatory, The University of Chicago.)

couldn't explain. At this point he abandoned Mercury and turned, with considerable success, to the computation of the orbits of comets.

In the summer of 1845, Le Verrier set out on his daunting quest for the masked perturber, completely unaware that Adams was close to a solution. As one of only a handful of mathematical astronomers with sufficient skill and patience to tackle the problem, Le Verrier was subject, like Adams, to the allure of discovering a new world by computation. Both men dreamed of demonstrating the power of their inscrutable mathematical science by pinpointing an unknown planet without once looking into the skies.

Le Verrier presented a preliminary report to the Paris Academy of Science in November of 1845, demonstrating mathematically that Uranus's wanderings were consistent enough to be caused by the planet's intrinsic motion and were not the result of faulty observations. Seven months later, he presented a second report to the academy, in which he calculated the planet's supposed position. But no one offered to search for it.

Meanwhile, still rebuffed by Airy and Challis, Adams continued to refine his calculations. Airy, for his part, began to take the whole matter more seriously when he saw that the estimate of the unknown planet's location presented in Le Verrier's second paper was remarkably close to the one computed by Adams, though the two had used somewhat different methods. Airy was sufficiently impressed to pass the news on to Challis and others, but still there was no immediate action.

It was not until the summer of 1846 that Challis finally mounted a search for Adams's unknown planet. Adams had provided updated position predictions and had even suggested that the planet would be large enough to appear as a disk, making it an inviting astronomical target. But Challis lacked confidence in Adams's mathematical machinations and began a systematic search across a wide swath of sky instead of concentrating on the small region that Adams had spotlighted. Like many others, Challis respected the practitioners of celestial mechanics, which seemed almost a black art, but he couldn't quite bring himself to trust the results.

Le Verrier also kept at it, presenting a third paper on the Uranus question at the end of August. Having further refined his calculations, he laid out what he believed were the orbital characteristics, mass, and position of the unknown disturber. He, too, predicted that the planet would be large enough to show up as a disk, adding that it would lie at a point near its closest approach to Earth and should be visible nearly all night. Academy members were impressed, but no one felt compelled to train a telescope on the designated spot in the sky.

Frustrated that none of his countrymen would search for the planet, Le Verrier finally wrote to young Johann Gottfried Galle, an assistant at the Berlin Observatory. A year earlier, Galle had sent Le Verrier a copy of his doctoral dissertation, and Le Verrier now took the opportunity to thank him and to praise his work. He also passed on his prediction. In this September 18, 1846, letter to Galle, Le Verrier explained: "At present, I am looking for a persistent observer, who is prepared to sacrifice some time to examining an area of the sky where there is possibly a new planet to be discovered. I came to this conclusion from our theory of Uranus. . . . It is impossible to account properly for observations of Uranus, unless the effect of a new, previously unknown, planet is introduced." He then added, "Direct your telescope to the point on the ecliptic in the constellation of Aquarius, in longitude 326 degrees, and you will find within a degree of that place a new planet, looking like a star of about the 9th magnitude, and having a perceptible disk."

Galle pleaded with the director of the Berlin Observatory to allow him to begin a search, and the director, who happened to be celebrating his birthday that night, consented to let his young assistant try his luck. On the night of September 23, 1846, Galle and graduate student Heinrich d'Arrest settled down for what they expected to be a long night of observing. Indeed, the first moments were discouraging. Looking in the direction specified by Le Verrier, Galle couldn't pick out anything planetlike among the myriad stars visible in the telescope. The two astronomers had to fall back on observing individual stars and checking them against standard star charts, a tedious chore made even more frustrating by inaccuracies in plotted star positions. Luckily, they came across an excellent new chart of the Aquarius region and began their search in earnest, with Galle calling out the position and brightness of each star in his field of view and d'Arrest noting whether the observations matched the chart.

In less than an hour, after examining only a handful of stars, they found one that was not on the chart and was no more than a degree away from Le Verrier's predicted position. Galle and d'Arrest believed they had found the planet, but they had to look again the next night to confirm that the object actually moved among the stars. Galle sent Le Verrier the news on the morning of September 25, writing gleefully, "The planet whose position you have pointed out actually exists." He concluded his letter by suggesting the new planet be named Janus. Le Verrier, annoyed by what he saw as an attempt to deny him his right, as its true discoverer, to name the planet, dubbed it Neptune. In his reply of thanks to Galle, he credited the French Bureau des Longi-

The discovery of Neptune aroused a great deal of public interest, but the controversy over who should properly get credit for finding the new planet at times received even more attention. This cartoon, published in the November 7, 1846, issue of *L'Illustration, Journal Universel,* presented the French viewpoint: that Adams had made his discovery by perusing Le Verrier's work. An issue of *Punch, or The London Charivari* on the same date published its own jaundiced view of the whole affair. It noted: "It is a great pity that when a thing of the kind turns up, the original finder is not able to mark it with his initials, or to take some course to prove his right to the article." (Library of Congress.)

tudes with officially bestowing this name on the planet, falsely invoking the bureau's prestige to head off Galle's proposal.

From that point on, the sensational news spread rapidly, arousing immense interest not only in astronomical circles but also among the educated public. The discovery of Neptune by mathematical calculation, as accounts at the time represented it, was considered a great triumph for Newtonian mechanics and for human reasoning. Of course, astronomers might have found the planet without relying on any calculations at all. But they had the whole sky to observe, with its thousands of stars, and their telescopes covered only tiny portions at any one time. Calculation, based on physical principles, was a shortcut to discovery.

Ironically, while the rest of the world marveled at this brilliant demonstration of scientific reasoning, the astronomical community had to face some disturbing facts. With Neptune identified, astronomers set about the significantly simpler, though still uncertain and laborious, task of computing its orbit from measurements of its position. This was made easier when it was learned, belatedly, that Challis had actually observed Neptune three times during his summer sweeps of the sky but had failed to recognize it as the planet of Adams's prediction.

The results of the calculations were surprising. It appeared that both Le Verrier and Adams, who had, respectively, placed Neptune within a degree and between 2 and 3 degrees of its observed position, had based their calculations on a faulty assumption. Neptune was actually much closer to the sun than either had guessed. Both men had started with a preliminary orbit at twice the distance of Uranus from the sun. But that figure was far from the true value of 1.57:1. Furthermore, using Kepler's third law, Adams had also determined that Neptune's orbital period should be 227 years, while its actual period is 165 years. If the orbit was nearly circular, a correct calculation of its position in 1840 would in 1780 have been wrong by 30 degrees. In addition, calculations of the gravitational force exerted by the unknown disturber depended on the choice of its mass. Adams had used a mass 45 times and Le Verrier a mass 32 times that of Earth. The true value, which emerged from Adams's later calculation of Neptune's orbit, was 17 times the mass of Earth.

Nonetheless, by a remarkable coincidence, Adams's method made up for these errors by positing a strongly eccentric orbit, with its nearest approach to the sun (its perihelion) occurring right when the planet was also closest to Uranus. Le Verrier's method produced a slightly different orbit, with similar compensating quirks that permitted Neptune's discovery. In other words, the places the real planet occupied in space and the gravitational force it exerted on Uranus differed widely from theoretical assumptions. With hindsight, it can be seen that the impressive calculations of Le Verrier and Adams merely cloaked fundamentally flawed assumptions, and their results were

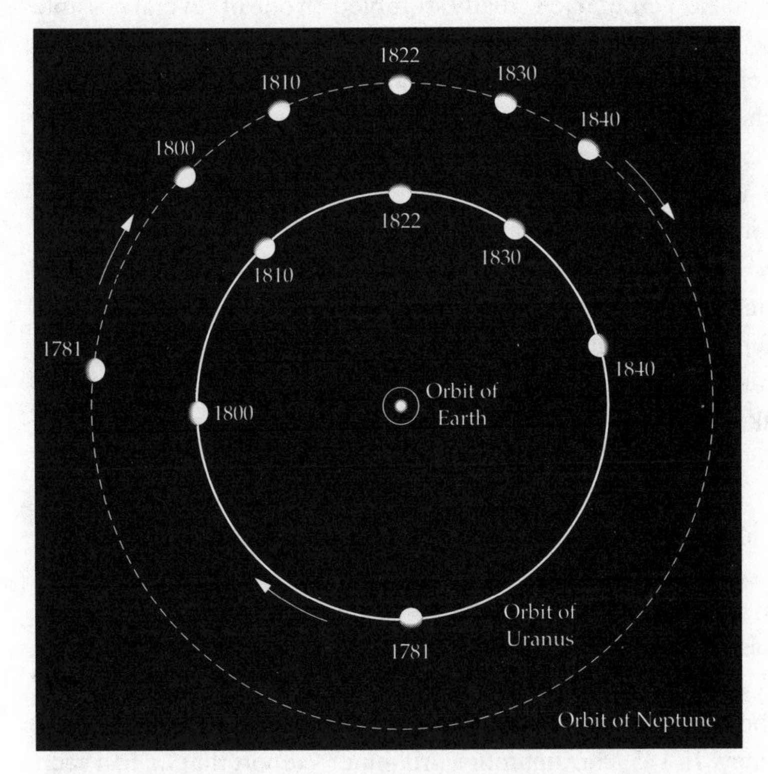

The relative positions of Uranus and Neptune in the years between 1781 and 1840. Before 1822, Neptune's pull made Uranus move faster along its orbit so that it reached positions ahead of expectations based on calculations taking into account the gravitational effects of the known planets. After 1822, Neptune's pull retarded Uranus.

really no better than their initial guesses of the planet's key parameters. Modern calculations demonstrate that Le Verrier's computations performed 40 years earlier or later would have produced a discrepancy between the predicted and actual position too great for the planet to have been discovered.

These flaws did not go unrecognized in the decades immediately following Neptune's discovery. Astronomers in the United States pointed out several of them and even showed that the inverse problem—using observed deviations in Uranus's orbit to infer the orbital elements of Neptune—actually produced more than one perfectly acceptable mathematical solution. By using the distances they did, Adams and Le Verrier had actually stumbled on one of several possible solutions, whereas the real Neptune followed a course corresponding to an alternative solution of the equations.

In the meantime, emboldened by his successful prediction and hoping that other planets could also be found by mathematical methods, Le Verrier spent the remaining 31 years of his life deeply immersed in the study of planetary perturbations. He was particularly bothered by his previous failure to account fully for the persistent discrepancy in Mercury's orbit. Astronomers knew that Mercury's perihelion advances along its orbit at a rate of 9 minutes and 26 seconds per century. In other words, Mercury's orbit slowly rotates in space, and it takes the orbit 227,000 years to return to its original position and orientation. By including the forces exerted by other planets, Le Verrier's calculations of 1843 had accounted for 90 percent of this motion, implying that the orbit should take 244,000 years to make one rotation. That left about 38 arc seconds (now known to be 43 arc seconds) per century unaccounted for by the gravity of known objects in the solar system.

Le Verrier came to believe that an unknown planet or a cluster of hidden bodies, closer to the sun than Mercury, was responsible for this discrepancy. In 1859, an amateur astronomer's report that he had seen a planet crossing the sun's face seemed to confirm this hypothesis, and Le Verrier was ready to claim that the object, which he named Vulcan, was the unknown planet of his calculations. But no one ever duplicated the finding, and Vulcan faded from the astronomical record.

In reality, rather than finding evidence of an inner planet, Le Verrier had unwittingly stumbled upon a situation in which Newton's

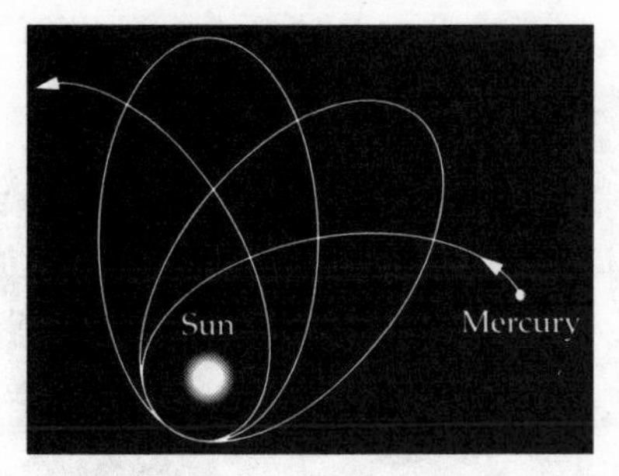

As Mercury races around the sun, its elliptical orbit slowly rotates. The actual effect, however, is greatly exaggerated in this diagram. Mercury's elliptical orbit takes about 227,000 years to complete a single revolution.

gravitational law failed to give sufficiently accurate results. That discrepancy was not resolved until Albert Einstein published his general theory of relativity in 1917, which precisely accounted for the extra 43 arc seconds per century in Mercury's orbital motion.

Einstein's general theory leads to a set of differential equations for planetary orbits identical to those of Newtonian theory except that one contains a small correction term. That modification of Newton's simple formula for gravitation depends on the square of the ratio of the planet's velocity and the speed of light. Only in the case of Mercury, the fastest moving planet, is this ratio large enough for its effect on an orbit's rotation to be readily discernible.

Of course, Mercury wasn't the only object of intense computational and telescopic scrutiny. There was no reason to suppose that Neptune was the solar system's outermost planet. Was there in fact a planet beyond Neptune, and could its presence be inferred from tiny irregularities in Neptune's motion? Neptune, however, crept along in its orbit so slowly that, even by the early twentieth century, its path was not known with sufficient precision to suggest a ninth, outlying planet.

That didn't stop William H. Pickering of the Harvard Observatory and Percival Lowell, eccentric businessman, astronomer, and mathematician (and founder of the Lowell Observatory in Flagstaff, Arizona), from trying. Both independently expended a great deal of effort sifting through the orbital irregularities of Uranus and Neptune for evidence

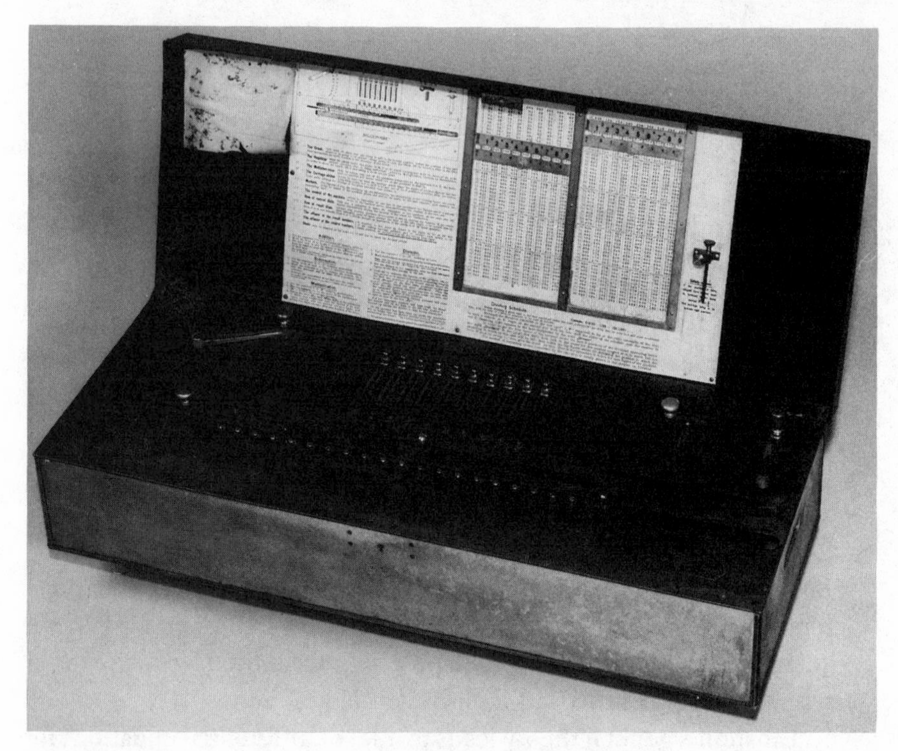

Introduced in 1893 for business applications, the "Millionaire" desktop calculator proved such a commercial success that nearly 5000 were sold between 1894 and 1935, with government agencies the biggest customers. This example was used at the U.S. Naval Observatory. (Smithsonian Institution.)

of a still more remote planet. Their task was far more complicated than that faced by Adams and Le Verrier, and the odds were stacked against them.

When Lowell began, in 1905, to search for his planet X beyond Neptune, he expected the necessary computations to take more than three years. To speed up the calculations, he decided to take advantage of mechanical calculators, which were just beginning to be recognized as a superior, though expensive, means of calculation in astronomy and other sciences. He bought a hefty desktop machine called the "Millionaire," which had been introduced in 1893 as a tool for big business. But Lowell's group of human calculators still ended up relying mainly on ordinary arithmetic and extensive tables to complete the huge number of calculations required.

In contrast, Pickering turned to an older, less computation-intensive method to come up with a prediction. To establish the existence and predict the location of the missing planet he had labeled O, he counted on a graphical plot of the observed but crudely measured discrepancies, or residuals, in the orbit of Uranus. Pickering offered his prediction in 1908, at which point he also asked Lowell for help in the search. But Lowell feared his rival's competition and refused.

Even with his mechanical calculator and team of talented human calculators, Lowell didn't produce a prediction for publication until 1915. In the meantime, he maintained a watch on the skies for any moving object that might turn out to be a planet at the fringes of the solar system, relying on photographic plates obtained at regular intervals to look for a slight shift in the position of any faint object from one plate to the next. But he had no success, and the project temporarily came to a halt soon after Lowell died in 1916.

When the search resumed in 1929, the Lowell Observatory had the advantage of a special wide-field camera and the keen eyes of a 22-year-old Kansas farm boy. Clyde W. Tombaugh went through 90 million star images captured on dozens of photographic plates before he finally glimpsed the ninth planet—a tiny, faint dot that shifted its position just enough from one plate to the next to stand out as a distant planet. The object, first identified by Tombaugh on February 14, 1930, from plates taken on January 23 and 29, was the planet Pluto.

But Pluto presented its own surprises. It showed no disk. Although it was found within 6 degrees of the spot that Lowell had pinpointed in his extensive computations, the planet was much too small to influence Neptune or Uranus as the calculations seemed to indicate. Moreover, Pluto's orbit defines a plane far more tilted with respect to Earth's orbital plane than any of the other planets. What agreement there was between Lowell's prediction and the planet's observed position was nothing more than a fluke. What may have seemed another triumph of celestial mechanics proved to be only an accident.

But once Pluto was found, astronomers could calculate a rough approximation of the planet's orbit and trace its location in previous years. It turned out that Pluto had shown up, unrecognized, in several of Lowell's old Flagstaff photographic plates, in images recorded on March 19 and April 7, 1915.

However, the puzzling, though minute, irregularities in the orbits of Neptune and Uranus were still unaccounted for. If Pluto wasn't responsible for them, was there still another unseen body gravitationally ruffling the frigid outer reaches of the solar system waiting to be discovered?

Although Neptune hasn't yet completed a full orbit since its discovery, observational accuracy has improved to the point that the deviations noted earlier can be attributed largely to uncertainty regarding the planet's position in the sky. Inaccuracies in the listed positions of reference stars in old star catalogs proved to be the main culprit. Moreover, when the *Voyager 2* spacecraft arrived at Neptune in 1989, planetary scientists were able to use the data obtained to establish that much of the remaining drift was simply a consequence of their hitherto incomplete knowledge of Neptune's orbit.

At the same time, bothersome discrepancies remain in Uranus's orbit, and whether the fault lies with the observations or with the inherent limitations of the computations is not clear. Present-day mathematical astronomers face the same kind of dilemma that confronted Bouvard, Airy, Adams, and Le Verrier before the discovery of Neptune, but the deviations they work with in the orbit of Uranus are now less than 1 arc second instead of 100 arc seconds.

Robert Harrington of the U.S. Naval Observatory was the modern-day counterpart of Adams and Le Verrier, but he could call upon computers to identify the likeliest candidates responsible for disturbing Uranus. To prepare for his search, he auditioned planet after hypothetical planet, entering hundreds of thousands of sizes and positions into a computer and calculating their effects on the orbits of Uranus and Neptune. His best candidate for explaining the discrepancies was a body with a mass three to five times Earth's, lying about three times as far away as Neptune from the sun and following an elliptical orbit tilted about 30 degrees to the plane of the solar system. Such a planet would take more than a thousand years to complete a circuit of the sun. Harrington's predicted location of the planet in the northern part of the constellation Centaurus, however, has so far drawn a blank. Indeed, extensive visual and infrared searches across much of the sky have turned up no object massive enough to qualify as the tenth planet. Yet Pluto and its companion in orbit, Charon, are much too small to account for the observed irregularities in Neptune's orbit.

These discrepancies suggest that there must be some other mass out there.

The mathematical complexities of deciphering the orbits of Uranus and Neptune, however, pale in comparison with those arising from the motion of our own familiar moon. It was in the moon's astonishingly complicated movements that astronomers first glimpsed dynamical chaos and truly began to learn the limits of mathematical prediction.

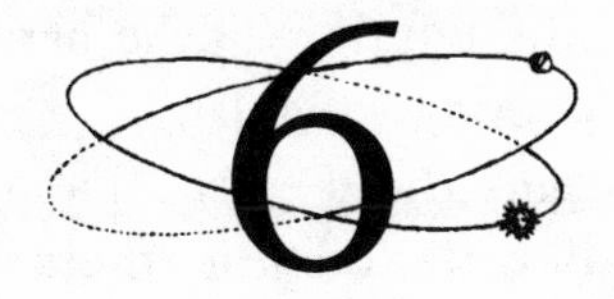

Inconstant Moon

•

The wandering moon,
Riding near her highest noon,
Like one that had been led astray
Through the heav'n's wide pathless way
JOHN MILTON (1608–1674), *Il Penseroso*

STONEHENGE STANDS as a solemn monument to predictability. Its prehistoric stone circle and huddled pillars, jutting from the Salisbury Plain in southern England, function as a kind of fossil celestial clock. The sight lines and alignments of this megalithic observatory mark key times in the rhythmic passage of the sun and moon across the sky and through the seasons. They testify to the power of the human mind to wrest a modicum of order from an unruly environment.

Four thousand years ago, inhabitants of the island now known as Britain were undoubtedly aware of the rising sun's cycle of excursions back and forth along the horizon. At the summer solstice in (by our calendar) late June, the sun would rise at its northernmost point along the horizon. At the winter solstice in late December, the sun would rise at its southernmost point. Each year at sunrise on successive days, the sun would step by step work its way to the same northernmost spot and then reverse direction, reaching its southernmost limit half a year later. Year after year, the plain dwellers would see the same pattern. Although the weather, the hunting, or the harvest might vary capriciously, the sun was reassuringly regular.

It's possible to imagine these early inhabitants starting to place markers pointing in the direction of the northern rising of the sun. Eventually, they would gain sufficient confidence in their data to erect a massive stone structure to commemorate their observations of the sun's motion. Indeed, the central avenue of Stonehenge points to the northernmost rising of the sun; still today, at the summer solstice, observers can see the rising sun dramatically skim a large, outlying marker—known as the Heel Stone—framed by the monument's ancient pillars.

The moon, too, marks the passage of time with its rhythms. Like the sun, it rises and sets at different locations along the horizon, but its cycle of excursions is much shorter. In fact, as described in Chapter 2, the moon completes nearly 13 such cycles in the year that it takes Earth to revolve once around the sun. The moon also periodically changes its appearance, waxing and waning through a cycle of phases ranging from crescent to full moon.

Constructed nearly 4000 years ago on Salisbury Plain in southern England, Stonehenge served as an astronomical monument. Alignments of various stones point to locations where the sun and moon rise at key times throughout the year.

It wouldn't be surprising to find that the ancestors of the builders of Stonehenge were also interested in determining the moon's northernmost rising location. But the moon's motion is far more complicated than the sun's. Instead of moonrise returning to the same northernmost point in each cycle, it shifts along the horizon from cycle to cycle. Only after nearly 19 years does moonrise occur again at roughly the same northernmost point. Early settlers of the region would have had to expend considerable effort patiently tracking the moon's position over many years to discern a clear pattern.

The evidence that these inhabitants may actually have tried to capture the moon's wanderings rests in curious formations that antedate the megalith we call Stonehenge. The mounds, ditches, postholes, and earthwork ring of this early lunar observatory apparently marked not only the moon's rising at the northernmost point in its 18.6-year cycle but also such moments as the midwinter rising of the full moon. Stonehenge itself, erected several hundred years later, appears to preserve some of these lunar alignments in the arrangement of its stones. Stonehenge, and particularly its less conspicuous, more humble predecessors on the same site, represents one of the first known attempts to bring some order to the moon's complex motion.

Since before recorded history, astronomers have tried to describe and predict the moon's path across the sky. Whether through stone monoliths or the abstract language of mathematics, countless hours and entire careers have been devoted to solving the mystery of the moon. Success, however, has proved elusive.

Thousands of years ago, observers didn't have the stopwatches or telescopes we now use to monitor the moon's movements moment by moment. But they had an abundance of lunar cycles to track and ponder. For instance, the time that elapses between successive new moons (or full moons) is about two days longer than the time it takes the moon to return to the same location relative to the stars. This means that the so-called synodic month—29.5 days long and associated with the moon's changing phases—is perpetually out of step with the sidereal month, which is 27.3 days long and associated with the moon's position relative to the stars.

Relying solely on naked-eye observations, early astronomers concentrated on the timing of particular celestial events, especially those that recur at regular intervals. They could see, for example, how the

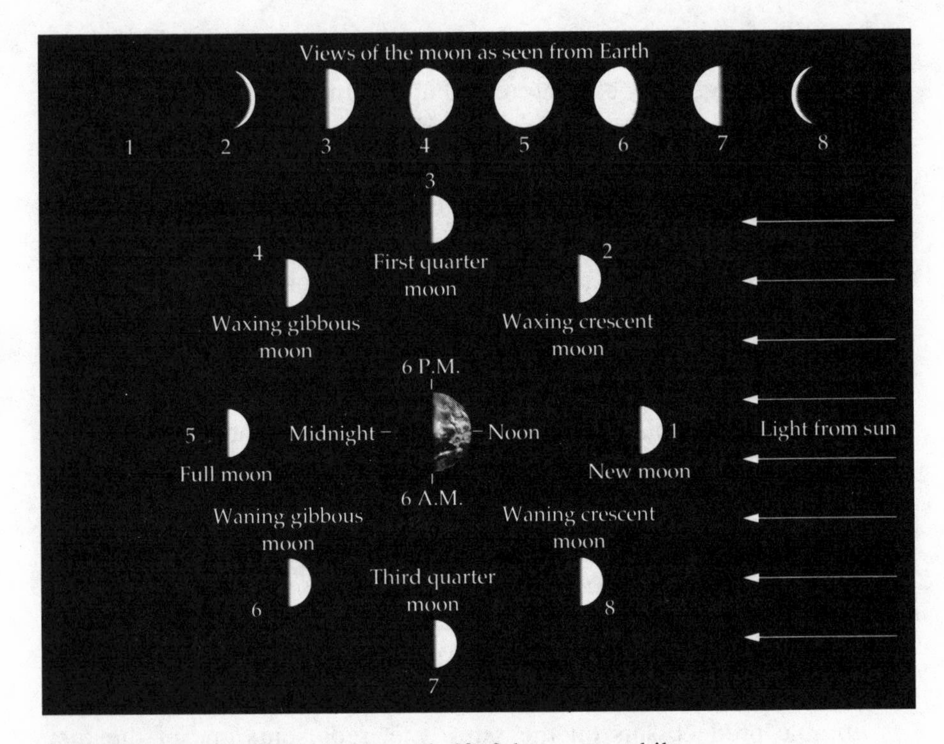

Light from the sun illuminates one half of the moon, while the other half is dark. As the moon orbits Earth, Earth-based observers see varying amounts of the moon's illuminated hemisphere. It takes 29 1/2 days for the moon to go through all its phases.

moon's monthly swings north and south of the sun's path, or ecliptic, affect its height in the sky and the times of moonrise and moonset throughout the year. But the periodic occurrence of eclipses, when the moon either passes in front of the sun or dips into Earth's shadow, probably made a much more profound impression.

The idea that a mythical dragon or other celestial monster was eating away at the full moon's glowing disk during a lunar eclipse no doubt inspired considerable wonder and fear in early human societies. In a culture with a literate and numerate elite, designated individuals charged with monitoring the sky for omens may well have recorded carefully the times and locations of successive eclipses. They would have noted, for instance, that lunar eclipses (in which the moon passes through Earth's shadow) occur only during a full moon. Furthermore, eclipses happen roughly twice a year, very close to when the moon's apparent path among the stars crosses the sun's. Suspecting a pattern in these events, ambitious rulers, priests, and politicians would have collected the data necessary to predict such threatening and dramatic events reliably, thereby impressing the populace with their power. But woe to the putative prognosticator who missed the dragon's meal by a day or two!

In the second century, the Alexandrian geographer and astronomer Ptolemy synthesized the insights and data of his astronomer predecessors into a single geometrical model of lunar motion that encompassed a variety of lunar cycles. For more than a thousand years, his model provided the mathematical framework for predicting the moon's position at key moments in its journey across the sky.

Imagine Earth at the center of a merry-go-round. In the simplest of all possible systems, the moon would sit on the merry-go-round's rim, tracing out a circle as it revolves around Earth. But to replicate the

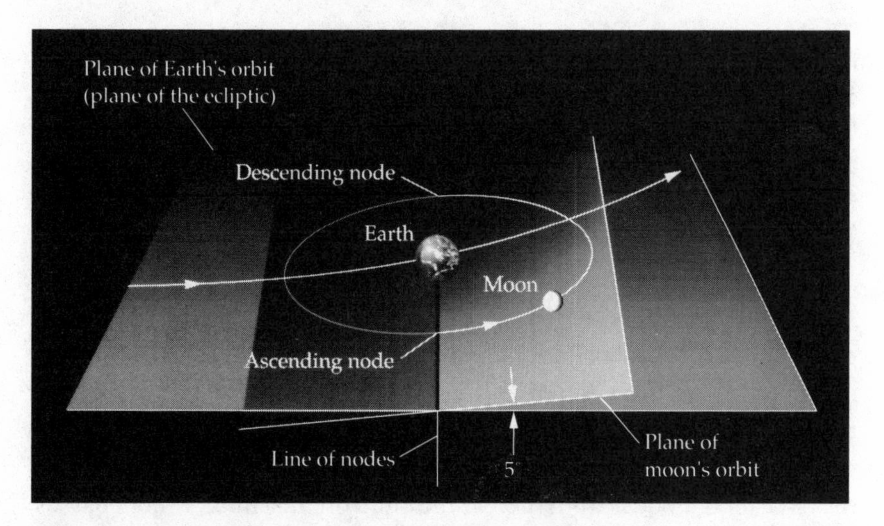

The plane of the moon's orbit is tilted about 5 degrees with respect to the plane of Earth's orbit. The line where the two planes intersect is called the line of nodes. A solar eclipse, in which the moon passes between the sun and Earth, occurs only if the moon is very near the line of nodes at new moon. A lunar eclipse, in which Earth's shadow obscures the moon's face, occurs only if the moon is very near the line of nodes at full moon.

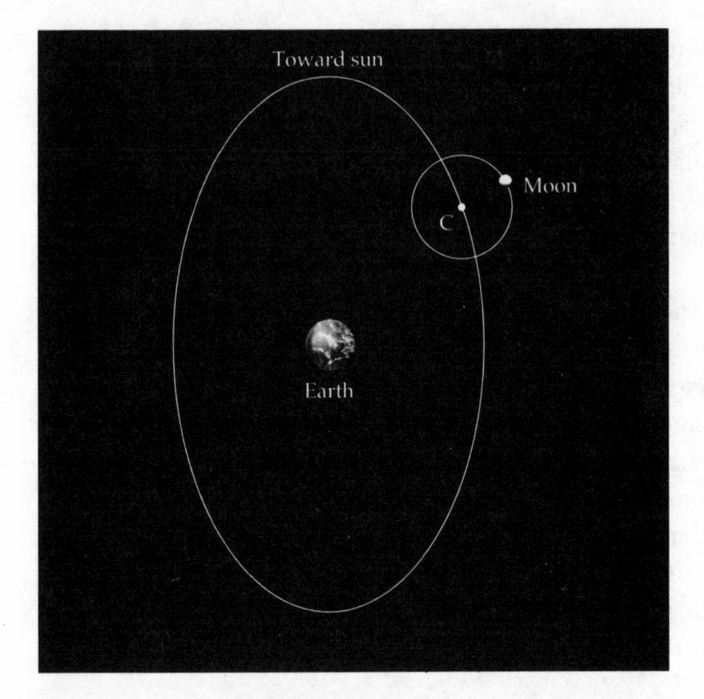

To obtain more accurate predictions of the moon's position in noneclipse situations, Ptolemy used what was, in effect, a highly elliptical orbit in addition to an epicycle centered at C to capture the moon's peregrinations. However, his mathematical solution suggested that the moon's apparent diameter, as observed from Earth, would change much more than observations indicated.

moon's true motion, as understood by Ptolemy, one must imagine the moon sitting on the edge of a tilted platform about 5 degrees steeper than the platform on which the sun rides around Earth. The line along which the imaginary platforms of the sun and moon intersect marks the so-called nodes, and the orientation of this line, which passes through Earth, slowly turns in a direction opposite to the circular motions of the sun and moon.

Now things really start to get complicated. Because its speed and distance from Earth actually vary, the moon itself rides on its own merry-go-round at the edge of the main, tilted platform. This smaller platform, about one-eleventh the radius of the larger one, also rotates slightly more slowly. This means that the location of the moon's closest approach to Earth, and hence its moment of highest speed, doesn't occur at the same point during each revolution of the main platform. Like the line joining the nodes, this point turns slowly but at a rate roughly twice as fast as the nodal line and in a direction opposite to that of the drifting nodes.

This was the essence of Ptolemy's array of tilted circles and epicycles devised to represent the moon's motion. But Ptolemy added one feature that earlier astronomers had failed to identify. His own observations, made with an instrument he had invented to determine the daytime latitude of the moon relative to the sun, had revealed that the moon's variations in speed were actually more complicated than suggested by eclipse data, which always caught the moon only when it was full. He found that the moon significantly speeds up and slows down during its monthly circuits, lagging or anticipating predicted positions at the first and last quarters, yet ending up at the predicted spot when it is new or full. This lunar excursion had gone unnoticed because previous observers had concentrated on lunar and solar eclipses rather than on direct observations of the moon's position.

Astronomers of succeeding generations gradually uncovered additional hitches in the moon's rhythms. These deviations from simple curves and uniform velocities persistently stymied the efforts of mathematical astronomers to develop a lunar theory that accounted completely for the moon's movements and allowed its position in the sky to be predicted accurately years into the future. Whether constructed from stone or mathematics, any apparatus designed to describe or replicate the moon's motion was necessarily extremely complicated.

By the sixteenth century, computation of the moon's position and the prediction of eclipses had become an important industry. Responding to the needs of navigators, calendar makers, and religious leaders, astrologer-mathematicians prepared and published tables of lunar and planetary positions for their patrons. But because the painstakingly prepared predictions inevitably drifted from observed positions, these tables had to be recomputed regularly. Moreover, different human calculators, using different mathematical schemes, made different errors and often produced discrepant results.

Such inconsistencies bothered Tycho Brahe, whose meticulous observations were destined to play a key role in Kepler's work (see Chapter 3). The renovation of the entire science of astronomy that Tycho had so long envisioned hinged on precise observation, yet the tangle of theory and data typical of astronomy in the sixteenth century prevented him from describing the moon's motion properly. Dissatisfied with the mathematical model developed by Copernicus, he tried to create an alternative approach, using eclipses as important checkpoints.

An eclipse was a major event at Tycho's observatory on the Danish island of Hven. He would mobilize all his assistants to ensure that it was well observed and precisely measured for its full duration. But because an eclipse lasts about two hours and predictions of its timing could err by as much as three hours, such an effort could occupy a considerable period of time.

In the first direct test of his own newly formulated theory of the moon's motion, Tycho used the moon's position on December 28,

DE NOVA STELLA ANNI 1572 67

haq; cùm Solis quasi socia sit, magnamq; cum eo familiaritatem & correspondentiam obtineat, tum in aliis, tum praesertim Motu & apparentiis: & insuper plurimum conducat, ejus motiones ad amussim perspectas habere, opera precium me facturum existimavi, si brevem & succinctam curriculi Lunaris restitutionem huic capiti de Sole, à quo etiam dependet, appendicis loco subjungerem; postquam multorum annorum accuratis observationibus satis exploratum habuerim, ejus in Caelo phaenomena non congruere hypothesibus hactenus constitutis, sive Ptolemaicis, sive Copernicais; atq; numeris hinc quomodocunq; derivatis: idq; non tantum secundum Longitudinem, sed & Latitudinem. In his enim utrisq; multo major invenitur inaequalitatis apparentis implicatio & varietas, quàm hactenus à quoquam animadversum est. Quod non saltem Eclipses ultra integram horam saepenumero consuetos canones eludentes (ut de earundem magnitudine, aliter subinde sese exhibente, nihil dicam) sed & majores adhuc in aliis locis incidentes disconvenientia evidenter testantur, veluti haec

ECLIPSES LUNÆ UNUM ET VIGINTI, SOLIS NOVEM A NOBIS DILIGENTER OBSERVATÆ.

ECLIPSES LUNÆ.

Anni	Men.	Dies	H. M.	Digi.
1573	Decem.	8	8 3	totalis
1576	Octob.	7	11 32	non pat:
1577	April.	2	8 50	totalis
1577	Septem.	26	13 3	totalis
1578	Septem.	15	13 17	2½
1580	Ianuar.	31	10 9	totalis
1581	Ianuar.	19	9 57	totalis
1581	Iul:	15	16 57	totalis
1584	Novem:	7	13 12	totalis
1587	Septem:	6	9 16	9 45
1588	Mart.	2	15 2	totalis
1590	Decem.	30	6 55	non pat:
1592	Iunij.	14	10 16	8 0
1592	Decem.	8	7 41	noa pat:
1594	Octob.	9	19 26	non pat:
1595	April.	13	16 36	totalis

ECLIPSES LUNÆ.

Anni	Men.	Dies	H. M.	Digi.
1595	Octob.	7	20 29	totalis
1596	April.	2	9 29	non pat:
1598	Feb:	9	18 7	11 30
1598	Augußt.	6	7 37	totalis
1599	Ianuar.	30	17 50	totalis

ECLIPSES SOLIS

Anni	Men.	Dies	H. M.	Digi.
1567	April.	9	0 0	6 20
1579	Febr.	25	5 50	5 50
1584	April.	30	5 39	3 0
1590	Iulij	21	7 54	5 0
1591	Iulij	10	3 33	2 30
1595	Septem:	23	1 8	3 50
1598	Febr.	24	23 16	9 20
1599	Iulij	11	16 8	3 0
1600	Iun:	30	1 44	5 0

& alia ex nostra restitutione, ubi cum ipso Caelo, aliorumq; calculis collata fuerit, patebunt. Hoc autem loco solùm modò ea, quae huc praecipuè conducunt, & maximè necessaria sunt, quàm brevissimè attingam, uberiorem de ☾ quoq; tractationem suo tempore, volente Numine, exhibiturus. Praemissa igitur praecedenti pagina observationes aliquot Eclipsium à me intra annos 27, quà fieri potuit diligentiâ, facta, quibus Lunaris motionis ratio majori ex parte, & quoad principaliora, innititur.

Tycho Brahe kept meticulous records of lunar and solar eclipses, as seen in these collected data. (Library of Congress.)

1590, to calculate that a lunar eclipse scheduled to occur two days later would begin at 6:24 P.M. Published predictions of other prognosticators put its starting time anywhere from 4:53 to 6:51 P.M. The eclipse began as Tycho and his household were still at supper, and by the time he and his assistants had assembled at 6:05 P.M., it was already half over. Tycho was in a foul mood for days afterward, and he temporarily abandoned his detailed study of the moon.

But the problem continued to vex him, and he returned to it again a few years later. Eventually he identified an additional lunar motion that had escaped Copernicus and other observers. He discovered that the inclination of the moon's orbit is not 5 degrees as previously assumed but oscillates between a minimum of 4 degrees 58.5 minutes at the full and new moons and a maximum of 5 degrees 17.5 minutes at the quarters. In other words, the platform defining the moon's motion around Earth is not only tilted but also wobbles slightly so that its slope periodically varies.

Tycho's obsession with monitoring the moon's course through all its phases, year after year, also revealed a variation in the moon's velocity when it is halfway between the quarter and full (or new) positions. In addition, he observed that the moon is consistently a little slower in winter and a little faster in summer than anyone had hitherto noted.

Quite independently and without taking measurements, Johannes Kepler had perceived the same annual deviation. Asked why a spring lunar eclipse in 1598 had occurred one and a half hours later than he had predicted in the provincial almanac, Kepler replied that the sun had a retarding influence on the moon, especially in winter when the sun is closest to Earth. This hypothesis was the first to suggest a cause for the moon's repeated wanderings from a simple path around Earth.

Many years later, after considerable thought and calculation, Kepler was able to construct a remarkably comprehensive lunar theory that substituted ellipses for the epicycles and eccentric circles of Ptolemy and Copernicus. His conceptually simple but mathematically sophisticated theory explicitly took into account the sun's influence on the moon's motion around Earth. He used this reformulated theory as the basis for calculating new lunar tables, which were used for decades after their completion in 1627.

It was Isaac Newton's theory of gravitation, as crystallized in the *Principia,* that clearly brought cause and effect into the picture and established an entirely new mathematical framework for both describing and explaining the moon's motion. In Newton's model, Ptolemy's multiple merry-go-round became an intricate, three-way tug-of-war, with Earth, moon, and sun attracting one another by means of a force determined only by their masses and distances apart. The complicated, periodic irregularities identified by Ptolemy and Tycho could now be attributed to the sun's disturbing influence on the moon's roughly elliptical orbit around Earth.

Calculating these effects so that they matched observation to within a few arc minutes, however, was beyond even Newton's capacities. Unhappy with his first stab at lunar theory in the *Principia,* he doggedly returned to the subject for nearly a year, beginning in the summer of 1694. The excruciating complexity of the mathematical problem, combined with his acrimonious, frustrating efforts to extract the lunar observations he needed from Astronomer Royal John Flamsteed at the Greenwich Observatory, took its toll. Newton later recalled bitterly that "his head never ached but with his studies on the moon." After nearly a year of intense labor, calculations based on his refined theory still erred by as much as 10 arc minutes from observed lunar positions.

Newton regarded his work on lunar theory as a great failure. Indeed, this final burst of concentrated cogitation marked the end of his scientific creativity. He devoted the remaining 34 years of his life to administrative affairs, though he took some time from his official duties to rework the results of earlier scientific and mathematical endeavors. In 1702, Newton consented to have his lunar theory published. In that edition, a note to the reader explains:

"The Irregularity of the Moon's Motion hath been all along the just Complaint of Astronomers; and indeed I have always look'd upon it as a great Misfortune that a Planet so near to us as the Moon is, and which might be so wonderfully useful to us by her Motion, as well as her Light and Attraction (by which our Tides are chiefly occasioned) should have her Orbit so unaccountably various, that it is in a manner vain to depend on any Calculation of an Ellipse, a Transit, or an Appulse of her, tho never so accurately made. Whereas could her place be but truly calculated, the Longitudes of Places would be found every where at Land with great Facility, and might be nearly guess'd at Sea without the help of a Telescope, which cannot there be used."

As Newton and other mathematicians and astronomers who followed in his footsteps found to their chagrin, the perturbing force exerted by the sun is a significant fraction of the total force acting on the moon, and simple perturbation theory can't fully account for the moon's motion. At the same time, because the moon is so near, it's easy to observe these small displacements. Often in the context of intense rivalries, different mathematicians adopted or invented somewhat different methods for solving problems in celestial mechanics. Each advance in the steady stream of refinements and extensions of calculus prompted fresh attacks on problems that had previously resisted solution. In contests driven as much by human weakness as by the pursuit of scientific or philosophical truth, the top mathematicians of the eighteenth century raced to solve the same basic problems in celestial mechanics. Books, papers, letters, and speeches at sessions of the scientific academies in Europe—particularly in London, Paris, and Berlin—resounded with claims and counterclaims. The promise of fame, influence, prestige, and sometimes remuneration fueled bitter disputes over the strengths and weaknesses of competing schemes. Because only a long series of accurate observations could settle such issues, arguments festered for decades.

Lacking ready-made formulas, adventurous theorists in the eighteenth century were forced to blaze new trails in a largely unexplored mathematical wilderness. Especially in lunar calculations, it was easy to get lost or end up on a tortuous mathematical path that consumed much time but afforded little illumination. As a result, choosing the appropriate course developed into a fine art practiced by an elite group of investigators.

One strategy was to derive from physical theory, with no additional input from observation, relationships that numerically linked the moon's characteristic periods or cycles. As he noted in the *Principia*, Newton was able, by means of certain approximations, to derive a simple algebraic expression linking the following four key lunar periods defined by Hipparchus 2000 years earlier:

Sidereal year, comprising 365.257 days, is the time it takes the sun to complete its yearly motion through the sky from a given position among the stars back to its original position.

Sidereal month, comprising 27.32166 days, is the time it takes the moon to move from a certain position among the stars back to its original position. The synodic month of 29.53059 days represents a compounded motion measured by the sidereal year combined with the sidereal month.

Anomalistic month, comprising 27.55455 days, is the time it takes the moon to complete its "anomaly," or cycle of speeding up and slowing down between one point of closest approach to Earth and the next.

Draconitic (named for the moon-eating dragon that causes eclipses), nodical, or tropical month, comprising 27.21222 days, is the time it takes the moon to move from one ascending or descending node to the next.

Newton's formula, expressed in terms of the ratio of the sidereal month to the sidereal year, gives the fraction by which the draconitic month is shorter than the sidereal month and establishes the 18.6-year cycle for the revolution of the nodes. According to Newton, this same formula also gives the fraction by which the anomalistic month is longer than the sidereal month. These spectacular results, derived entirely from theory, encapsulate in a wonderfully compact manner a significant insight into the workings of the solar system; namely, why the moon's motion has certain well-defined rhythms.

The formula's simplicity, however, is marred by the fact that the second fraction, related to the motion of the moon's perigee (its point of nearest approach to Earth), gives a result for the anomalistic period that is about twice as long as the observed value. Recent studies of Newton's unpublished papers reveal that he actually carried out a second treatment of the problem, which gives the correct period for the anomalistic motion. Apparently aware that he had failed to obtain the right answer the first time around, Newton reworked his data to get the result he wanted; however, the theoretical problem remained unresolved. In subsequent years,

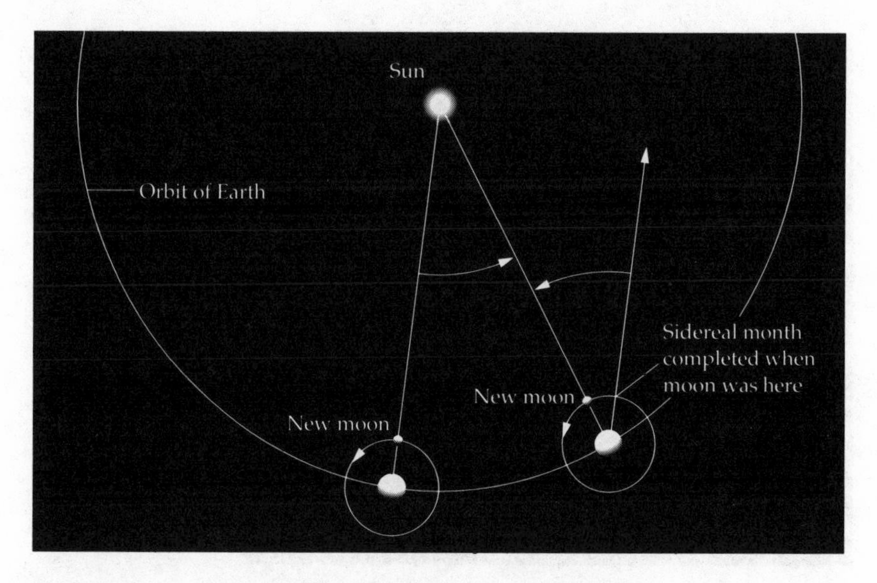

The sidereal month represents the time it takes the moon to complete one full revolution around Earth with respect to the background stars. However, because Earth is constantly moving along its orbit about the sun, the moon must travel through slightly more than 360 degrees to get from one new moon to the next. Thus the synodic month, which is linked to the phases of the moon, is slightly longer than the sidereal month.

it became a goal of mathematicians to improve upon Newton's flawed result, and the English government and several scientific societies in Europe offered prizes for a calculation that resolved the celebrated discrepancy.

Among those attracted to the problem were the prominent mathematicians Alexis Claude Clairaut, Jean Le Rond d'Alembert, and Leonhard Euler. Where had Newton gone wrong? Initially, all three independently concluded that Newton's law of gravitation could not satisfactorily explain the motion of the lunar orbit's perigee (or its apogee). In a letter to Clairaut, dated September 30, 1747, Euler wrote, "I am able to give several proofs that the forces which act on the moon do not exactly follow the rule of Newton, and the one you draw from the movement of the apogee is the most striking, and I have clearly pointed this out in my lunar theory. . . . Since the errors cannot be attributed to the observations, I do not doubt

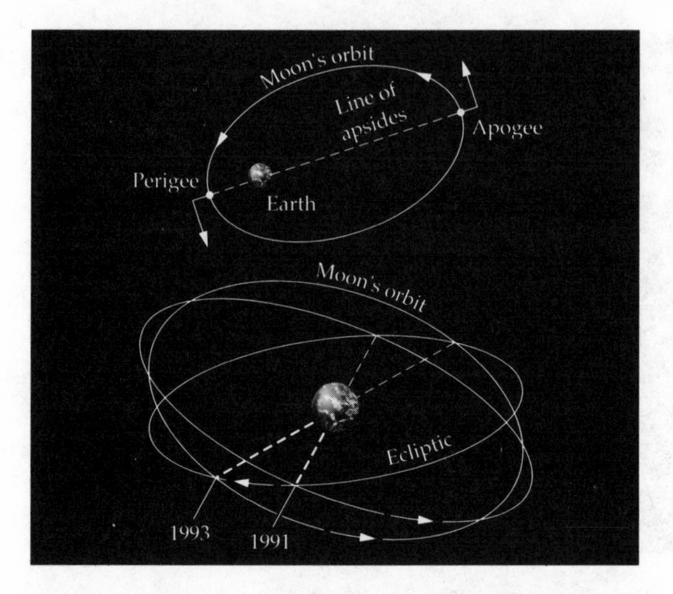

The moon follows an elliptical orbit around Earth. It is at perigee when it is nearest Earth and at apogee when it is farthest from Earth. The line connecting the points of perigee and apogee is called the line of apsides. The sun's gravitational pull on the moon as it moves in its orbit around Earth causes both the line of nodes and the line of apsides to shift slowly over the years. The line of nodes gradually moves westward, taking 18.61 years to complete one full rotation *(bottom)*. While the line of nodes shifts westward with respect to the constellations, the sun's gravity causes the line of apsides to move eastward *(top)*. It takes 8.85 years for the line of apsides to complete one full rotation

that a certain derangement of the forces supposed in the theory is the cause."

Aware of, and no doubt encouraged by, Euler's opinion, Clairaut went before the Academy of Science in Paris to announce confidently that his own work showed the inverse-square law of gravitation to be inadequate for describing the moon's motion. D'Alembert, who was working on the same problem and getting similar results, kept a low profile, except in correspondence with a few colleagues. On June 16, 1748, he wrote with characteristic caution, "I plan to publish next year and perhaps at the beginning of the year all my researches on these

Left: Jean Le Rond d'Alembert (1717–1783). *Right:* Alexis Claude Clairaut (1713–1765). (Smithsonian Institution.)

things . . . but I am so much afraid of making assertions on such an important matter that I am in no hurry to publish anything on the subject. Besides I will be very sorry to overthrow Newton."

Clairaut's announcement before the Academy of Science represented the opening salvo in a battle for recognition marked by secret research, malicious gossip, hasty announcements, unfortunate errors, and public rancor. D'Alembert and Clairaut, in particular, developed an intense rivalry that lasted for decades and eventually degenerated into open hostility. Their clashes were often discussed in the fashionable salons of Paris, and over the years, despite the esoteric nature of the debates in the academy, the passion and vitriol of the combatants and their supporters attracted large audiences.

Even among the educated and interested, few could follow the technical arguments. Commenting on these debates, the eighteenth-century philosopher Denis Diderot wrote of the mathematicians that they "resemble those who gaze out from the tops of high mountains whose summits are lost in the clouds. Objects on the plain below have disappeared from view; they are left with only the spectacle of their own thoughts and the consciousness of the height to which they have risen and where perhaps it is not possible for everyone to follow and breathe [the thin air]."

Ironically, d'Alembert's initial caution was rewarded when Clairaut was forced to retract his original paper and announce on May 17, 1749, that Newton's law of gravitation had been correct all along. D'Alembert himself found that he had misread a sign in his lunar tables, and the resulting calculations had been flawed. Both mathematicians, using different methods, eventually resolved the problem that had stymied Newton. Clairaut, in particular, discovered that by including terms that had previously been neglected, he could bring Newtonian theory into agreement with observation. In other words, Newton's formula was just the first term in a series. In the case of the anomalistic cycle, the second term contributed enough to modify significantly an answer based only on the first term.

Although Clairaut and d'Alembert resolved the problem to their satisfaction, Euler still could not find the error in his own lunar theory. And because Clairaut's announcement had not included the method used to come up with the final result, Euler had no clues to work with. Desperate to see the details of Clairaut's method, he contrived to have the St. Petersburg Academy award its prize for 1752 to the best paper on the anomalistic motion of the moon's orbit. His stratagem worked. Clairaut couldn't resist sending in an essay, with the help of which Euler was able to uncover the error in his own work. As a result, Euler succeeded in finally bringing his theory into agreement with Clairaut's conclusions. D'Alembert had also considered entering the contest but, believing that Euler would use his considerable influence with the academy to award the prize to Clairaut, decided to have his theory published separately. In the end, d'Alembert's paper appeared about nine months before the publication of Clairaut's essay and the accompanying lunar tables.

At this stage, the battleground shifted to the computation of lunar tables, and here d'Alembert was at a disadvantage. Knowing nothing about observational astronomy and depending wholly on an analytical approach, he struggled to draw up tables of sufficient accuracy to predict the moon's position in future years.

Clairaut, who took a more pragmatic approach by blending theory with observational data, dismissed d'Alembert's efforts as having little

Newton's formulas describing the moon's motion

Newton attempted to derive an expression describing the moon's motion in terms of four periods characteristic of its movements. In his derivation based on gravitational theory and the laws of motion, he linked the sidereal year (T_0 = 365.257 days), the sidereal month (T_1 = 27.32166 days), the anomalistic month (T_2 = 27.55455 days), and the draconitic month (T_3 = 27.21222 days), showing that in terms of the ratio $m = T_1/T_0$, one gets the formulas $T_2/T_1 - 1 = +3m^2/4$ and $T_3/T_1 - 1 = -3m^2/4$. However, although the second formula gives roughly the correct answer when appropriate values are substituted into the expression, the first one is off by a factor of 2.

Many years later, Clairaut and d'Alembert were able to show that Newton's expressions were merely the initial terms in a so-called power series, expressed in terms of powers of m:

$$\frac{T_2}{T_1} - 1 = +3\frac{m^2}{4} + 225\frac{m^3}{32} + \ldots$$

$$\frac{T_3}{T_1} - 1 = -3\frac{m^2}{4} + 9\frac{m^3}{32} - \ldots$$

In the second case, the addition of the second term changes the overall answer relatively little, whereas in the first case the second term is large enough to affect the sum significantly.

practical value. Conceding that d'Alembert's mathematical methods were more elegant and probably superior to his own, he nonetheless bitterly attacked their purely theoretical basis. In 1758, Clairaut felt compelled to defend his own approach and the quality of his lunar tables by heaping ridicule on the theorists, particularly d'Alembert: "In order to avoid delicate experiments or long and tedious calculations, in order to substitute analytical methods which cost them less trouble, they often make hypotheses which have no place in nature; they pursue theories that are foreign to their object, whereas a little constancy in the execution of a purely simple method would surely have brought them to their goal."

Indeed, the best tables available in the late eighteenth and early nineteenth centuries for navigation and other practical purposes were those of Tobias Mayer, who had combined the most important perturbations from Euler's lunar theory with a large amount of observational data. By limiting computations to no more than 14 terms and using adjustable parameters to fudge the theory, he could obtain results that differed by as little as 1.5 arc minutes from actual lunar positions. That uncertainty amounted to only a few dozen miles, a relatively insignificant distance when ascertaining positions far out at sea.

To d'Alembert, such makeshift tables were obviously inferior to those derived from theory. He believed that accuracy was less important than the refinement of mathematical methods. This argument over theory and practice highlighted a growing division between experimental and theoretical scientists. That division persists to this day in the computer-intensive methods used to calculate nautical tables and spacecraft trajectories, versus the study of long-term trends in the moon's motion, where theory plays a significant role.

Perfection, both practical and theoretical, remained out of reach throughout the eighteenth and into the nineteenth century. Perturbation theory could indicate where the deviations lay, what their periods were, and how they depended on the sun, the nodes, and the perigee. The computation of sufficiently exact amounts for the perturbations, however, proved elusive. After a lifelong struggle, Euler threw up his hands in defeat. In the preface of his great work on lunar theory, he wrote, "As often as I have tried these forty years to derive the theory and motion of the moon from the principles of gravitation, there always arose so many difficulties that I am compelled to break off my work and latest researches. The problem reduces to three differential equations of the second degree, which not only cannot be integrated in any way but which also put the greatest difficulties in the way of approximations with which we must here content ourselves; so that I do not see how, by means of theory alone, this research can be completed, nay, not even solely adapted to any useful purpose."

Later mathematicians, such as Joseph-Louis Lagrange and Pierre-Simon de Laplace, also devoted a great deal of time to lunar theory. They tried various strategies, reformulating Newton's equations to emphasize different variables or to focus on such conserved quantities as energy and angular momentum. In the process, they discovered a lot of new mathematics valuable not only in celestial mechanics but

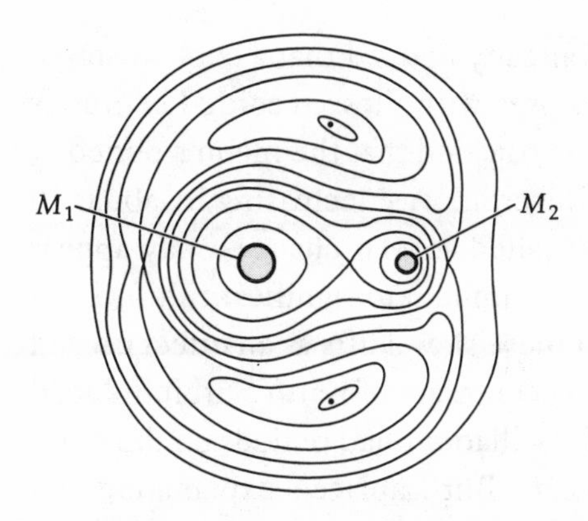

In the late 1700s, Joseph Louis Lagrange considered a special case of the three-body problem, in which one of the objects is small enough that its gravity has virtually no effect on the other two, more massive objects. Solutions of this "restricted" three-body problem serve as a starting point for tackling the moon's movements and provide a means of calculating the course of a spaceship from Earth to the moon or the path of an asteroid affected mainly by the pulls of the sun and Jupiter. This diagram shows one way of displaying the combined gravitational fields of two massive objects. Each line represents an equipotential contour, along which a small object would feel the same gravitational force. The resulting contour map serves as a sort of topographic map showing "hills" and "valleys" in the gravitational field. Any small object in this field will feel a force pulling it in the "downhill" direction. In this example, the mass of M_1 is greater than that of M_2.

also in engineering physics, a burgeoning field in the wake of the great construction projects and mechanical inventions of the industrial revolution.

Using a purely theoretical approach, Laplace succeeded in representing the moon's movements to within less than 0.5 arc minute, or 30 arc seconds. His lunar theory included terms that depend on such factors as the size of Earth's equatorial bulge and the ratio of the distances of the sun and moon to Earth. He also noted a mysterious perturbation that appeared to correspond not to an oscillation but to a long-term trend, which would bring the moon very slowly, but steadily, closer to Earth.

Edmond Halley had in 1693 already noticed that a comparison of contemporary observations of eclipses with those recorded centuries earlier in Arabic and other sources indicated that the moon's period of rotation, and hence the radius of its orbit, was decreasing by about 10 arc seconds per century. Such so-called secular changes may appear small, but over millions of years they amount to significant shifts in an orbit. In 1787, Laplace explained these slow shifts as an effect caused by long-period oscillations in the eccentricity of Earth's orbit induced by the other planets. These small oscillations had periods on the order of several tens of thousands of years. But Laplace's explanation was only half true. Even taking this effect into account, there remained an apparent secular effect, but one that indicated an increase in the lunar orbital period.

Thus, the moon consistently eluded the ingenious geometrical and numerical snares set by these theorists. Even in the nineteenth century, mathematical astronomers could do no better with their mighty calculations based on Newton's mechanics than to match the accuracy of lunar periods computed from eclipse records by Greek and Babylonian astronomers more than 2000 years earlier. At the same time, new instruments capable of refined astronomical measurements revealed additional deviations in the moon's motion that also fell outside contemporary theory.

Near the end of the nineteenth century, George William Hill, an astronomer at the U.S. Naval Almanac Office, discovered a computational trick that considerably simplified the mathematical machinery of lunar theory. Whereas Lagrange and others typically began with ellipses and then modified these simple orbits step by step to accommodate the effect of a third body, Hill started with a particular orbit defined by a special, simple solution of the three-body problem. It exploited the fact that although mathematicians could not come up with a general formula describing the motions of all three bodies for all time, they could find precise solutions for certain special cases. Hill's starting point was a particular periodic orbit that already included the sun's perturbing influence. He then mathematically superimposed additional wiggles and shifts representing the movements of the lunar perigee and nodes to bring this main, smooth loop closer to the moon's true orbit.

By taking such a course, Hill considerably reduced the labor involved in computing lunar orbits and positions to a given precision,

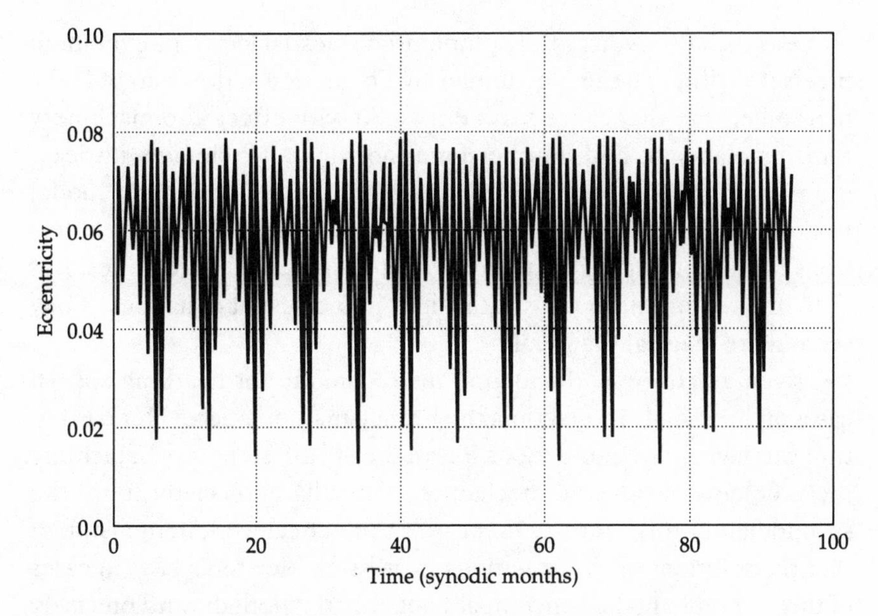

The complexity and subtle irregularity of the moon's motion show up in this plot of the eccentricity of the lunar trajectory over time, measured in synodic months starting in the year 1980. (Reprinted with permission from Martin C. Gutzwiller, *Chaos in Classical and Quantum Mechanics.* New York: Springer-Verlag, 1990.)

because the successive terms in the algebraic series expressing his lunar theory decreased in value rapidly. One didn't have to compute a large number of terms before reaching an acceptable answer. Indeed, Hill's method worked not only for the moon but also for any three-body problem involving planetary satellites or a trio of stars. Suitably modified and refined by Ernest W. Brown and others, it formed the basis for all subsequent lunar calculations and, in effect, helped land the first men on the moon. Although the advent of electronic digital computers in the 1950s didn't simplify the computations further, it did make the calculation of such orbits much faster and more convenient.

In the development of lunar theory, nearly 2000 years separate Hill and Brown from Hipparchus and Ptolemy. Newton's arguments succeeded only partially in bringing theory into agreement with observation. Subsequent refinements led to abstruse calculations that obscured rather than illuminated the reasons that the moon and planets move as they do, even as they pinpointed celestial locations ever more precisely. Although the Newtonian laws of motion at the heart of Hill's method ensure that cause stays entwined with effect, the machinery remains elaborate and cumbersome, not unlike Ptolemy's intricate merry-go-round of circular orbits and epicycles. Ptolemy's model worked surprisingly well, as do today's complicated algebraic expressions for computing planetary motions and spacecraft trajectories. But they often afford little physical insight into celestial dynamics. They seem more magical than real.

Even as Hill was elucidating his technique for handling special cases of three-body interaction, the French mathematician Henri Poincaré was laying the foundations for a radically different way of tackling the problems of celestial mechanics. Not fully appreciated until the second half of this century, Poincaré's theoretical work demonstrated that the solutions of the equations expressing Newton's laws in cases of three or more bodies encompass not only the periodic and precisely predictable but also the irregular and unpredictable. In the words of Martin C. Gutzwiller, who at one time was involved in the intricacies of computing lunar movements and now pursues chaos in the still stranger realm of quantum mechanics, "The moon's motion is only a mild case of a congenital disease" afflicting the mathematical description of nearly all physical systems.

Prophet of Chaos

•

Without dimension, where length,
breadth, and height,
And time and place are lost; where eldest night
And chaos, ancestors of nature, hold
Eternal anarchy

JOHN MILTON (1608–1674), *Paradise Lost*

FOR THE most part, mathematicians are no more immune to the lure of contests and prizes than are today's entertainers or athletes. Indeed, the tradition of offering substantial rewards for solutions of important mathematical problems goes back many centuries. From the time of Newton, mathematicians eagerly vied for honors offered by scientific societies, governments, and an assortment of academies and special-interest groups. Such prizes recognized not only significant achievements but also future promise, and young scholars strove to impress or shake the scientific establishment while venerable authorities maneuvered to maintain the status quo. Highly political at times, these competitions also engendered international quarrels, bitter personal disputes, and intense rivalries.

When Gösta Mittag-Leffler, a mathematics professor at the fledgling school that eventually grew into the University of Stockholm, proposed a contest as part of the sixtieth-anniversary celebration of the birth of Oscar II, King of Sweden and Norway, he no doubt expected that a certain amount of intrigue would accompany the awarding of the prize on January 21, 1889. What he couldn't foresee were the ensuing scandal and the monumental, seminal piece of mathematical work that would emerge from the controversy and mark the true birth of the mathematical study of what we now call dynamical chaos.

For the contest, mathematicians were invited to write an original paper addressing one of four questions. Proposed largely by Karl Weierstrass, who had been Mittag-Leffler's teacher at the University of Berlin, these questions highlighted a number of key issues at the frontiers of mathematical research.

At that time, Weierstrass was well established as a central figure in mathematics. With his commitment to clear thinking and to the systematic development of mathematical concepts, he was a significant force in a field that was striving to pull itself free of the physical world and to redefine itself with new rigor and increasing abstraction. His masterly lectures at the University of Berlin regularly attracted enthusiastic crowds of mathematicians and students from all over the world.

Left: Gösta Mittag-Leffler (1846–1927). *Right:* Karl Weierstrass (1815–1897). (Courtesy of the Mittag-Leffler Institute.)

Methodical and painstaking, Weierstrass published his results reluctantly and only after extensive and thorough revisions. He distrusted intuition and took considerable delight in disdainfully pointing out flaws in purported mathematical proofs put forward by others. Not surprisingly, the numerous victims of his barbed comments and criticisms regarded him as insufferably arrogant.

One of Weierstrass's questions concerned celestial mechanics. It arose out of a seemingly innocent but suggestive remark made nearly 30 years earlier by Peter Gustav Lejeune Dirichlet, a prominent mathematician at the University of Göttingen. In 1858, Dirichlet had told his student Leopold Kronecker that he had discovered a novel method of solving certain differential equations. Dirichlet hinted that by applying this method to the equations of celestial mechanics, he could prove in an absolutely rigorous fashion that the solar system, as modeled by those equations, is stable.

The mathematical question of the solar system's stability hinges on the nature of the solutions of the differential equations used to describe planetary motions. When astronomers or mathematicians derived these solutions in terms of series—the sums arising from infinitely long chains of algebraic terms—they counted on the fact that they could obtain reasonable estimates of planetary behavior by evaluating just the first few terms of such series. They assumed that the remaining terms mattered very little to the final answer.

The validity of such a procedure depended on the nature of the series involved. Mathematicians already knew that certain series converge; that is, step-by-step evaluation of the terms in such series produces answers closer and closer to a single, well-defined value. Other series diverge. In these cases, the inclusion of additional terms leads to larger and larger sums, which never approach a particular value.

It wasn't clear, however, into which category the particular series generated in celestial mechanics by the application of perturbation theory—series like those developed by Charles Delaunay in his lunar theory (see Chapter 9)—would fall. If someone could prove that these series converge, then it would also be possible to demonstrate that the solar system they described is stable. Such a proof would help settle the question of whether planets continue repeating the same, basic movements forever or could radically change their orbits. It was conceivable, for example, that even the relatively weak forces between

Infinite series

Infinite series play an important role in mathematics and its applications in science, including celestial mechanics. The following string of fractions is one example of an infinite series:

$$1 + \frac{1}{2} + \frac{1}{3} + \frac{1}{4} + \frac{1}{5} + \cdots$$

Some series obviously get bigger and bigger as one includes more and more terms. Consider, for example, this series:

$$1 + 2 + 4 + 8 + 16 + \cdots$$

This particular series tends to infinity. Adding more terms simply makes it get bigger at a faster rate. In other words, this series diverges.

The first example given of an infinite series also approaches infinity, albeit much more slowly than the second example. One wouldn't initially suspect such a result from adding fractions that steadily get smaller and smaller, but the fractions prove large enough to make the series diverge.

On the other hand, certain infinite series converge to specific values. Evaluating additional terms in this particular infinite series brings its total closer and closer to the number $\pi/4$:

$$1 - \frac{1}{3} + \frac{1}{5} - \frac{1}{7} \cdots$$

Adding a million terms and multiplying that sum by 4 yields the result that $\pi = 3.1415937$, which is correct for the first five decimal places. One would need to evaluate even more terms to get a better value of π. In many cases, without a mathematical proof, it's difficult to tell whether a particular infinite series diverges to infinity or converges to a specific numerical value.

One can also express an infinite series in terms of a variable x to produce a so-called power series:

$$x - \frac{1}{2}x^2 + \frac{1}{3}x^3 - \frac{1}{4}x^4 \cdots$$

Isaac Newton was among the first to realize that many mathematical relationships can be expressed in terms of infinite series, and he learned to handle them in ingenious ways to solve mathematical and scientific problems. Such expressions also played an important role in the development of his version of what came to be called the calculus.

In celestial mechanics, approximate solutions of the appropriate equations of motion yield infinite series expressed in terms of such variables as an orbit's eccentricity or some other orbital parameters. Mathematical astronomers evaluate such expressions to as many terms as they believe necessary to make predictions of a certain accuracy. In some instances, however, they have no proof that the series they used actually converges to a specific value.

the planets could, over a sufficiently long time, shuffle planetary orbits and completely alter the solar system's arrangement. But it was no simple matter to extract from the differential equations that had been used since Newton's time to account for the movements of celestial bodies those interactions between individual planets that slowly but steadily accumulate to change the basic shape of their roughly elliptical orbits.

Dirichlet died a year after dropping his incendiary hint, and he left behind no written evidence of his alleged discovery. But Kronecker transmitted Dirichlet's tantalizing comment to the mathematical community. Because Dirichlet's known proofs had earned a reputation for sparkling elegance and diamond-cut rigor, his remark was taken quite seriously. Weierstrass, among others, put considerable effort into trying to recover this lost treasure, but all such attempts failed.

Stymied in his search, Weierstrass proposed as one of the prize problems precisely what he believed Dirichlet had achieved: "For a system of arbitrarily many mass points that attract each other according to Newton's laws, assuming that no two points ever collide, give the coordinates of the individual points for all time as the sum of a uniformly convergent series whose terms are made up of known functions." He went on to state quite bluntly his belief that there had to be a simple answer: "This problem, whose solution would considerably extend our understanding of the solar system, would seem capable of being solved using analytic methods presently at our disposal. . . . Unfortunately, we know nothing about [Dirichlet's] method. . . . We can nevertheless suppose, almost with certainty, that this method was based not on long and complicated calculations, but on the development of a fundamental and simple idea that one could reasonably hope to recover through persevering and penetrating research."

Baldly stated in its purely mathematical form, the problem represented an important theoretical issue in celestial mechanics. It also highlighted the steadily growing division during the nineteenth century between the application of mathematics to the solution of practical problems and the exploration of mathematics itself. The distant future of an ideal, arbitrarily populated solar system, in which planets are reduced to featureless points wandering in abstract three-dimensional space, held little meaning for navigators or almanac calculators. They wanted an accurate prediction of Jupiter's position on a certain date or the precise instant of the next lunar eclipse. In contrast, theorists in search of unsuspected mathematical subtleties had absolutely no interest in computing the moon's position to 20 decimal places, but they had a great deal of interest in how the mathematical machinery used to make such predictions really worked.

Assembling a jury of distinguished mathematicians to judge the expected entries presented its own difficulties. Mittag-Leffler wanted to ensure the prestige of the prize by convening the best possible international panel. But no one could imagine a group made up of such leading mathematicians as Weierstrass, Arthur Cayley of Cambridge, Charles Hermite of France, and Pafnuty Chebyshev of Russia ever arriving at a consensus on the merits of any submitted paper. Indeed, resentful at being lumped together, each of these aging luminaries would likely refuse to serve on a jury that included the other three. Weierstrass himself worried about reaching an agreement if the panel members were not able to meet face to face and had to rely on the mails for communicating their opinions.

As the months passed and King Oscar's birthday neared, the controversy over the composition of the jury grew. In the spring of 1885 Kronecker, at that time a colleague of Weierstrass in Berlin, injected himself into the politicking. Gravely offended that Weierstrass had been asked to propose the four prize questions and that, to make matters worse, Weierstrass had selected a question concerning certain algebraic relations on which Kronecker believed himself the foremost, if not the only, expert, Kronecker sought a way to get back at his rival. He wrote an angry letter to Mittag-Leffler threatening to inform the king that he had long ago proved the impossibility of obtaining the results that this particular problem demanded.

Mittag-Leffler tried to smooth the matter over. Adopting a conciliatory tone, he explained to Kronecker that he had simply wanted to honor Weierstrass because of his advanced age. But when news of that explanation reached Weierstrass, who was about to celebrate his seventieth birthday, he in turn was offended. Nonetheless, despite wounded egos and bad feeling, the project continued, and Kronecker's intrigues eventually produced a jury consisting of himself, Weierstrass, and Mittag-Leffler. The four problems, as posed by Weierstrass, remained unaltered.

The heavily promoted contest, with its substantial prize of a gold medal and the princely sum of 2500 crowns, attracted wide attention and an array of entries. Among these was a lengthy contribution from Jules Henri Poincaré, who only a few years earlier, at the age of 27, had become a professor at the prestigious University of Paris. Bearded, bespectacled, and absent-minded, he was already living up to the stereotype of a mathematician.

Equipped with an incredibly acute memory, Poincaré worked intuitively, often figuring out problems in his head as he restlessly paced back and forth in his quarters. Only after he had worked out an idea in his mind would he commit it to paper, and his written work nearly always showed signs of hasty composition. Poincaré intensely disliked retracing his steps to fill in gaps and to tidy up his reasoning and mathematical language. He preferred to leave such niceties to others. Satisfied that he had surmounted the crucial barriers and hacked out a rough path, he would plunge headlong into the next daunting thicket.

Poincaré's distinctive approach to mathematics was evident from the start. In 1875, at the age of 21, he had entered the School of Mines planning to become an engineer, but he spent nearly all his spare time developing a novel approach to differential equations. These investigations proved so fruitful that three years later he was able to present his results as a doctoral thesis at the University of Paris. Poincaré's thesis adviser, Gaston Darboux, was later to recall: "At first glance, it seemed clear to me that [Poincaré's thesis] was out of the ordinary and fully deserved to be accepted. Certainly it contained enough results to furnish material for several good theses. But it must be said without hesitation if an accurate idea is to be given of the way in which Poincaré worked, many points required corrections and explanations. . . . He willingly did the corrections and tidying which seemed necessary to me. But when I asked him to do so, he explained to me that he had many other ideas in his head; he was already occupied with some of the great problems whose solution he was to give us."

As outlined in Chapter 4, differential equations represent reality as a continuum that changes smoothly from one moment to the next. In essence, these expressions state the relationships that must hold between the values of such variables as position and velocity just a

Henri Poincaré (1854–1912). (American Institute of Physics.)

vanishingly small interval before and after a given moment. As Poincaré put it, "Instead of considering in its entirety the progressive development of a phenomenon, one simply seeks to relate one instant to the instant immediately preceding; one supposes that the actual state of the world depends only on the most recent past, without being directly influenced, so to speak, by the memory of the distant past. Thanks to this postulate, rather than studying directly a phenomenon's whole succession, one can limit oneself to writing out its 'differential equation'."

The problem remained of piecing together, or integrating, these infinitesimal segments to deduce from the expressed relationships the course of a phenomenon from a given initial state to a final state. More often than not, these equations proved difficult if not impossible to solve, or integrate, and practitioners of the differential art tended to focus on the minority of equations most amenable to attack. These expert mathematicians, scientists, and engineers usually looked at specific cases, considering only one set of possibilities at a time.

In contrast, Poincaré wanted to see the entire realm of possibilities at once. Influenced by George Hill's work on the three-body problem and lunar theory (mentioned in the previous chapter), Poincaré was interested in testing the general assumption that small changes in some parameter in a differential equation would result in only small differences in the numbers constituting the solution—that no slight change in a parameter would ever lead to a gross change in overall behavior. He also believed that his unique approach would eventually shed light on specific problems in celestial mechanics.

In the introduction to a collection of his early papers on differential equations, Poincaré remarked, "Could one not ask whether one of the bodies will always remain in a certain region of the heavens, or if it could just as well travel further and further away forever; whether the distance between two bodies will grow or diminish in the infinite future, or if it instead remains bracketed between certain limits forever? Could one not ask a thousand questions of this kind, which would all be solved once one understood how to construct qualitatively the trajectories of three bodies?"

Poincaré's strategy resembled that of a military commander warily venturing into unknown territory and sending out scouts along specific routes to ascertain the lay of the land. The territory traversed by Poincaré in his dynamical reconnaissance missions consisted of a peculiar, multidimensional construct known as phase space, which serves as the backdrop for the geometrical shapes and flows representing the totality of solutions of differential equations. The notion of phase space first arose in a novel, highly fruitful reformulation of Newtonian mechanics by the Irish mathematician William Rowan Hamilton. As Roger Penrose noted in his provocative critique of artificial intelligence, *The Emperor's New Mind*, "The form of the Hamiltonian equations allows us to 'visualize' the evolution of a classical system in a very powerful and general way."

Until Hamilton introduced his scheme, the positions of particles held center stage in mechanics, with velocity being merely the rate of change of position with respect to time. Hamilton's formulation shifted the focus from the velocities to the momenta of particles (the momentum of a particle being its mass multiplied by its velocity). This simple change had a significant impact on the solution of problems in mechanics. Because position and momentum can be treated as independent quantities on a more or less equal footing, two sets of differential equations can be used. One set tells how the momenta of various particles are changing over time, and the other describes how their positions are changing over time. All of these equations are derived from an expression that gives the total energy of a system, called the Hamiltonian function, in terms of all the position and momentum variables. In Penrose's words, "The Hamiltonian formulation provides a very elegant and symmetrical description of mechanics."

In this description, each particle making up a physical system has three momentum and three position coordinates, one for each of the three independent directions in space. Thus it takes six numbers, or coordinates, to specify the "state" of a single particle at any given instant, twelve to specify the states of two particles, and so on. One can then conceive of an abstract "space" of a large number of dimensions, with one dimension for each of the coordinates describing a physical system comprising a certain number of particles.

This mathematical construction is known as phase space, and it provides a powerful means of turning numbers into a kind of contoured road map of the allowed possibilities. In effect, no matter how complicated the system under consideration may be, a single point in phase space encapsulates the entire system's state of motion at a particular instant. Hamilton's differential equations give the rates of change of each of the coordinates, which in turn tell us how all the individual particles move. These expressions provide a direction of travel—an arrow—for each particle at any instant. Added together, these arrows supply the direction associated with a given point in phase space. The course of the resultant arrow over time describes the evolution of the entire system: as the state changes, so the point traces out a path through phase space. It's a matter of following the arrows. Indeed, solving the equations is equivalent to constructing curves of the arrows'

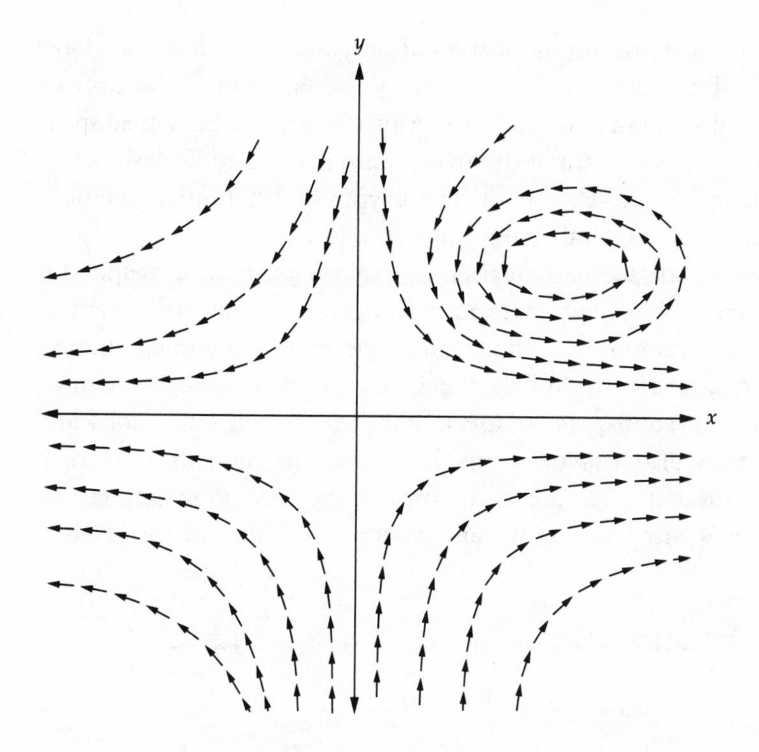

One way to represent the overall behavior of a set of differential equations is to draw little arrows, or vectors, indicating the direction of the "flow" at various points in space.

directions at every point. The result is a "trajectory" or "orbit" that traverses an abstract, essentially unimaginable space quite different in character from the three-dimensional space through which actual planets wander.

In such a geometrical approach, one can imagine all the possible solutions of a particular set of differential equations as a body of flowing water, with the equations specifying the velocity at every point in the stream. A small stick dropped into the water traces out a flow line as it wanders over the surface, sometimes floating swiftly along a simple, direct path, at other times getting caught in a languid swirl. When two sticks are dropped in succession at the same place, one can see whether they follow virtually identical trajectories and can explore the general pattern of flows in various neighborhoods. In this way, instead of focusing on specific, numerical solutions, one can obtain a global picture of the dynamics represented by the equation. Like drifting flecks of foam, such a portrait drawn in an abstract, mathematical space provides a vivid snapshot of the river's currents and eddies. Similarly, one can compare and classify different types of differential equations by looking at their overall geometries and flow patterns.

In phase space, mathematicians can visualize as a shape this whole range of behaviors packed into a given differential equation. For a simple system, that shape might be merely a curved surface, like a torus, which resembles the surface of a doughnut. In more complicated systems, the shape might span many dimensions and show an array of twists and turns. Each point on the surface of such a shape, whether a simple curve or a convoluted form, represents the system's state at an instant frozen in time. As the system

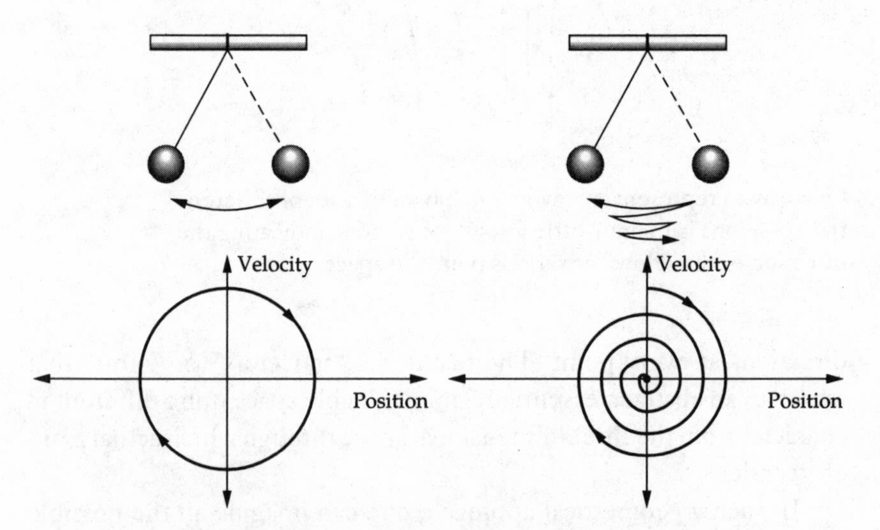

Phase space, an abstract space whose coordinates are position and velocity, is a useful and convenient concept for visualizing the behavior of a dynamical system. The motion of a pendulum *(top)*, for example, is completely determined by its initial position and velocity. As the pendulum swings back and forth, its motion appears as an "orbit" in phase space. For an ideal, frictionless pendulum, the orbit is a closed curve *(bottom left)*. With friction, the orbit spirals to a point *(bottom right)*. In other words, the pendulum comes to a stop.

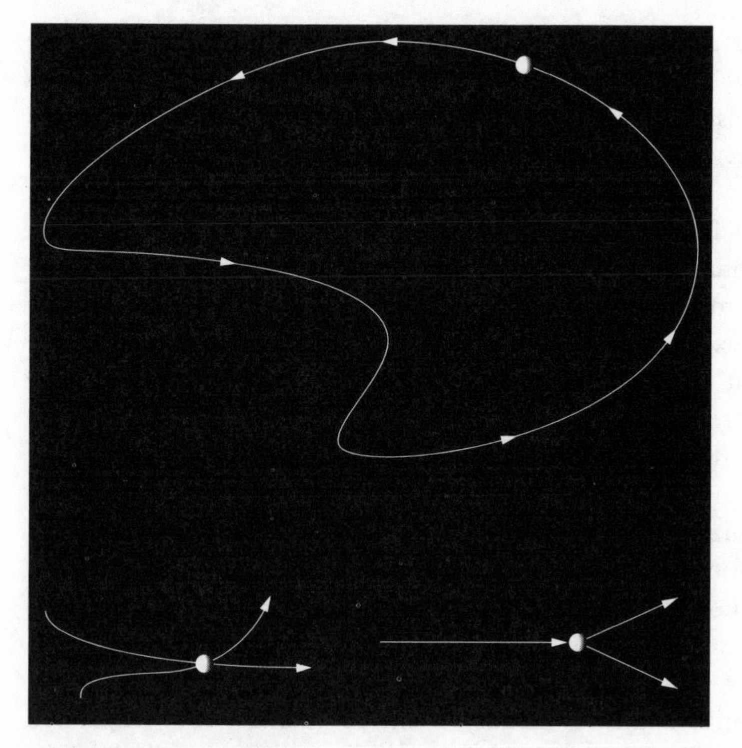

Any point in phase space that traces out a closed loop is destined to repeat the same motion again and again *(top)*. Trajectories in phase space, however, cannot intersect or branch *(bottom)*.

progresses through time the point moves, tracing out a trajectory across the surface. Changing a parameter in the differential equation bends the shape, but shapes that look roughly the same exhibit roughly the same kinds of behavior.

In the case of two gravitationally interacting bodies, solving the appropriate equations produces simple loops along a particular surface in phase space. When the motions represented by these loops are reinterpreted in physical space, they correspond to the ellipses discovered by Kepler. The addition of a third body, however, disturbs this appealingly simple picture. Now the energy and momentum conservation laws are no longer sufficient to restrict the possible motions to a tractable, easy-to-define surface or shape. Even when the problem is imagined as a case in which one of the bodies has so small a mass that it doesn't affect the other two—almost a two-body universe—the phase-space geometry still becomes remarkably complicated and difficult to visualize. Trajectories representing the histories of the three bodies have considerable freedom to wander and wind through phase space.

It was Poincaré's unique geometrical perspective that prompted him to enter Mittag-Leffler's contest and tackle the behavior of the solutions to the differential equations of celestial mechanics. Faced with the extreme difficulty of properly interpreting the interminable series that arise as solutions, he shifted the emphasis from specific solutions to their totality as represented by flows. To do this, he advanced the study of differential equations from numbers, formulas, and the manipulation of algebraic expressions to geometry, curves, and the visualization of flows. Instead of looking at the contents of his mathematical package, Poincaré looked at the package itself to obtain the clues he needed to determine whether a series converged and what the implications of that result were for the stability of a dynamical system.

Inventive and revolutionary, Poincaré's contest entry explained and then used geometrical ideas to characterize the dynamical behavior of three gravitationally interacting bodies. He examined specifically the so-called "restricted" three-body problem, which Hill had used to such great advantage in arriving at improved approximations of the moon's orbit. Poincaré's complex paper, more than 200 pages long, brought the contest's judges, especially Weierstrass, into unfamiliar mathematical territory. They needed time to sort through Poincaré's complicated reasoning, but they were also under great pressure to complete their deliberations in time for the celebrations marking King Oscar II's birthday. Finally, the pressure to make an announcement overruled whatever reservations the ailing, weary Weierstrass may have had. Poincaré's paper was clearly the best submission and deserved the prize. Weierstrass reported to Mittag-Leffler, "You may tell your sovereign that this work cannot indeed be considered as furnishing the complete solution of the question proposed, but that it is nevertheless of such importance that its publication will inaugurate a new era in the history of celestial mechanics. The end that His Majesty had in view in opening the competition may therefore be considered as having been attained."

Poincaré was declared the winner, and his prize-winning paper was published in Mittag-Leffler's increasingly influential journal, *Acta Mathematica,* to coincide with the birthday festivities. But doubts about Poincaré's proof surfaced soon after. In particular, Edvard Phragmén, one of Mittag-Leffler's colleagues, pointed out a serious error in the way Poincaré had interpreted, to prove stability, his geometrical picture of the requisite differential equations. News of the error prompted immediate complaints of unfair treatment accorded another mathematician and astronomer, who had submitted a paper on the same problem and used more conventional methods to prove convergence. Weierstrass himself later wrote to a colleague that he had noticed some possible errors in Poincaré's work but had been unable to correct them. Instead, he had submitted some cautionary notes that were meant to accompany Poincaré's paper in *Acta Mathematica.*

Deeply disturbed that such accusations would cast a pall over the prize and sully the reputations of all involved, Mittag-Leffler took the drastic measure of seizing all copies of the journal issue featuring Poincaré's paper. Instead of publishing a correction or admission of error in a later issue, he persuaded Poincaré to revise his proof and submit a new paper, which would then be published as the prize-winning work. By relentlessly tracking down and destroying all copies of the offending issue, Mittag-Leffler hoped to forestall further controversy and shut down discussion of the matter.

Poincaré himself made no attempt to hide the fact that he had made an error, though he never specified precisely what it was. Indeed, he took pains in the preface of his revised paper to acknowledge Phragmén's role in pointing it out. After a few years of relatively quiet grumbling and periodic efforts by certain individuals in the mathematics community to revive doubts about the merits of Poincaré's paper, Mittag-Leffler's cover-up in the end succeeded. The furor died down, and the entire incident gradually faded from memory. Only one complete copy of the journal issue containing Poincaré's original paper remains, and it rests in a locked drawer at the Mittag-Leffler Institute in Djursholm, a small town just a short distance northeast of Stockholm.

For Poincaré, the months between the prize ceremony and the publication of the revised paper in 1890 were a period of intense cogitation and frantic activity. He had to go back over his reasoning to

Introduction.

Le travail qui va suivre et qui a pour objet l'étude du problème des trois corps est un remaniement du mémoire que j'avais présenté au Concours pour le prix institué par Sa Majesté le Roi de Suède. Ce remaniement était devenu nécessaire pour plusieurs raisons. Pressé par le temps, j'avais dû énoncer quelques résultats sans démonstration; le lecteur n'aurait pu, à l'aide des indications que je donnais, reconstituer les démonstrations qu'avec beaucoup de peine. J'avais songé d'abord à publier le texte primitif en l'accompagnant de notes explicatives; mais j'avais été amené à multiplier ces notes de telle sorte que la lecture du mémoire serait devenue fastidieuse et pénible.

J'ai donc préféré fondre ces notes dans le corps de l'ouvrage, ce qui a l'avantage d'éviter quelques redites et de faire mieux ressortir l'ordre logique des idées.

Je dois beaucoup de reconnaissance a M. Phragmén qui non seulement a revu les épreuves avec beaucoup de soin, mais qui, ayant lu le mémoire avec attention et en ayant pénétré le sens avec une grande finesse, m'a signalé les points où des explications complémentaires lui semblaient nécessaires pour faciliter l'entière intelligence de ma pensée. Je lui dois la forme élégante que je donne au calcul de S_i^m et de T_i^m à la fin du § 12. C'est même lui qui, en appelant mon attention sur un point délicat, m'a permis de découvrir et de rectifier une importante erreur.

Dans quelques-unes des additions que j'ai faites au mémoire primitif, je me borne à rappeler certains résultats déjà connus; comme ces résultats sont dispersés dans un grand nombre de recueils et que j'en fais un fréquent usage, j'ai cru rendre service au lecteur en lui épargnant de fastidieuses recherches; d'ailleurs je suis souvent conduit à appliquer ces théorèmes sous une forme différente de celle que leur auteur leur avait d'abord donnée et il était indispensable de les exposer sous cette nouvelle forme. Ces théorèmes acquis, dont quelques-uns sont même classiques

In the introduction of his revised paper, Poincaré thanked Edvard Phragmén for so carefully reviewing his original proof. He went on, "It is also he who, in calling my attention to a delicate point, made it possible for me to discover and correct an important error."

clarify his geometrical interpretations of differential equations and the consequences of his results for celestial mechanics. Indeed, it was Poincaré's unique genius that allowed him to acknowledge, albeit reluctantly, that there was room for the unpredictable in deterministic systems.

Poincaré's revolutionary resolution of these thorny issues finally appeared in a 270-page memoir titled *Sur le problème des trois corps et les équations de la dynamique* (On the problem of three bodies and the equations of dynamics). His reworked paper, published in *Acta Mathematica*, proved a mathematical landmark, foreshadowing much contemporary research in dynamical systems. Any reader who dared to brave its difficult language, unconventional strategies, and daunting intricacies experienced a mind-jostling but richly rewarding ride into a new mathematics with startling implications for the mathematical modeling of physical phenomena. With devastating effect, Poincaré systematically demolished the structure that Weierstrass held so dear. He introduced doubt and uncertainty where the older man had anticipated a clean mathematical solution that would pave the way to perpetual certainty.

Poincaré began by establishing that although the equations representing three gravitationally interacting bodies yield a well-defined relationship between time and position, there exists no all-purpose, computational shortcut—no magic formula—for making accurate predictions of position far into the future. In other words, the series that arise out of perturbation theory typically diverge. Indeed, there was plenty of room for the unpredictable in a Newtonian system, and the question of stability could not be settled directly by examining the divergent series associated with solutions of the equations of motion for the solar system.

Nonetheless, although the three-body problem has no complete solution expressible in a compact form, one can obtain approximate solutions to practically any degree of accuracy. This means that computing the initial members of a series, expressed in terms of some measurable variable, provides satisfactory answers for a range of practical applications. That's what those interested in computing planetary and lunar positions had been doing for centuries and continue to do to this day.

Poincaré paved the way for a new kind of model that indicates the range of possibilities the future holds in store, but doesn't predict specifically which one will occur. Shattering the rigid framework imposed by the refined but limited tools of quantitative mathematics, he took the more indirect, qualitative approach of drawing pictures rather than performing calculations. Though less precise, the geometrical methods he deployed so skillfully had greater potential for revealing the future than conventional methods.

It was in these investigations that Poincaré first caught a glimpse of, and to some degree appreciated, what we now know as dynamical chaos. Ironically, it was the incorrect use of these novel methods that sabotaged his initial conclusions concerning the stability of the solar system. In his revised paper, he realized that these geometrical portraits revealed not stability but a bewildering dynamical domain of wondrous complexity.

In drawing his imaginary portraits, Poincaré was interested not in finding a formula for a particular solution of the equations of motion but in seeing how everything fitted together. To do this, he focused on those trajectories in phase space that repeat themselves at regular intervals. Such periodic orbits, after wandering about in phase space for a certain amount of time, always return precisely to their starting points before setting off again on exactly the same path. This cycle of positions and velocities repeats itself forever.

To find these loops in phase space, Poincaré chose to examine not the entire space and the families of geometrical surfaces defining all possible trajectories, but a cross section through that surface. In effect, he turned the differential equations describing the motion as a continuous flow into a set of steps specifying what happens to the motion at regularly spaced time intervals.

Mathematicians call the result of this procedure an iterated mapping. It's like taking a series of stroboscopic photographs that capture a movement at regular intervals. If the motion repeats itself exactly and the flashes occur at the appropriate frequency, the moving object appears at the same place, with the same velocity, every time. The corresponding Poincaré map, or cross section of phase space, shows a single point. If the motion is irregular, then each flash catches the object in a different location. In this case, the corresponding Poincaré map shows a sequence of points spread out across the map. However, unlike a true stroboscopic photograph, these points do not represent the actual positions of a body or group of bodies caught in a flashlamp's glare (except under special circumstances). Rather, they are an abstract representation of the totality of its motion.

One way to examine trajectories in phase space is to see what pattern they create when they puncture an imaginary sheet of paper placed in the way. The pattern displayed on such a "Poincaré section" provides a shorthand method of identifying various types of orbits in phase space. For example, a single point on such a diagram indicates a purely periodic motion *(top)*. A more complicated, cyclic motion that eventually repeats itself after four passages would show four distinct points *(bottom)*.

Poincaré discovered that he could obtain stroboscopic maps that show a single point or succession of points that eventually return to a certain position and repeat the cycle. Such instances correspond to periodic solutions of the underlying differential equations. He then focused on what happens to orbits in the immediate neighborhood of periodic orbits. He found that trajectories with slightly different starting points can separate amazingly quickly, rather than remain close together at all times. He also found that the resulting sequences of points tend to fill whole regions of the cross section, hinting at an underlying trajectory that wanders haphazardly through phase space, apparently never to return precisely to its starting point.

It was these strange, wandering trajectories that Poincaré initially overlooked and then caught a glimpse of when he revised his prize-winning paper. He later remarked in his monumental three-volume treatise *Les Méthodes nouvelles de la Mécanique céleste* (New methods in celestial mechanics), "One is struck by the complexity of this picture, which I do not even attempt to draw. Nothing can give us a better idea of the complexity of the three-body problem, and more generally of all the problems in dynamics . . . ."

Modern computer simulations reveal what Poincaré suspected but was not able to visualize in any direct way. Built up point by point, these diagrams suggest entire worlds explored by a single, meandering trajectory—an intricate landscape of islands, straits, mountains, and continents. As the trajectory wanders about, adding a spot to the map with each stroboscopic flash, some areas fill out faster than others while some remain altogether bare. Magnifying any part of this curious landscape reveals a hierarchical structure of great complexity: miniature islands breaking the surface in the straits between islands, tinier islets nestling in the narrow straits between the miniature islands, and so on, ad infinitum.

Because a single trajectory, starting at one particular point, can wander throughout a portion of phase space, it is also difficult to distinguish orbits with very long periods from those that never strictly repeat. Indeed, the underlying differential equations yield two radically different types of behavior: regular (periodic or almost periodic motion) and irregular (chaotic, nonperiodic but bounded, motion). Yet the global picture of their dynamics mixes order and randomness so intimately that it's impossible to tell where one ends and the other begins.

Of course, Poincaré didn't have sophisticated computational and graphics capabilities to enable him to see all this. His glimpses of this startling realm came indirectly through his theoretical considerations rather than through any numerical simulations. What a rare genius he must have been to envision this strange dynamical behavior that we now see so readily with the help of computational paraphernalia!

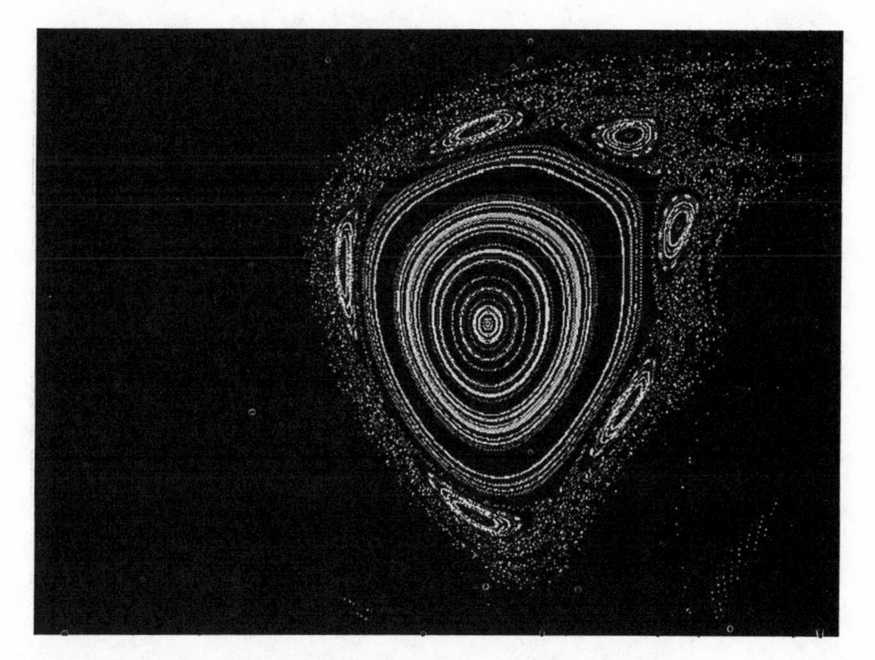

In many instances, phase-space trajectories can be so convoluted that the resulting Poincaré map shows an exceedingly complex pattern, featuring islands of regularity in a sea of chaos, which corresponds to unstable orbits that scatter points randomly.

Poincaré at first believed that he could settle the stability question by examining the patterns in his stroboscopic maps. What he detected were specific points where two surfaces appeared to intersect. Given his considerable experience in drawing and interpreting phase portraits, with their flow lines representing the behavior packed into a differential equation, Poincaré naturally but mistakenly applied some of the same notions to his analysis of the corresponding stroboscopic maps. Having already proved that flow lines in phase portraits can't cross, he assumed that the same rule, and whatever consequences followed, also held true for his maps. He interpreted the apparent crossings as the coming together of surfaces representing stable and unstable behavior, which thereby form some sort of barrier that confines the system's behavior and keeps it from wandering too far afield.

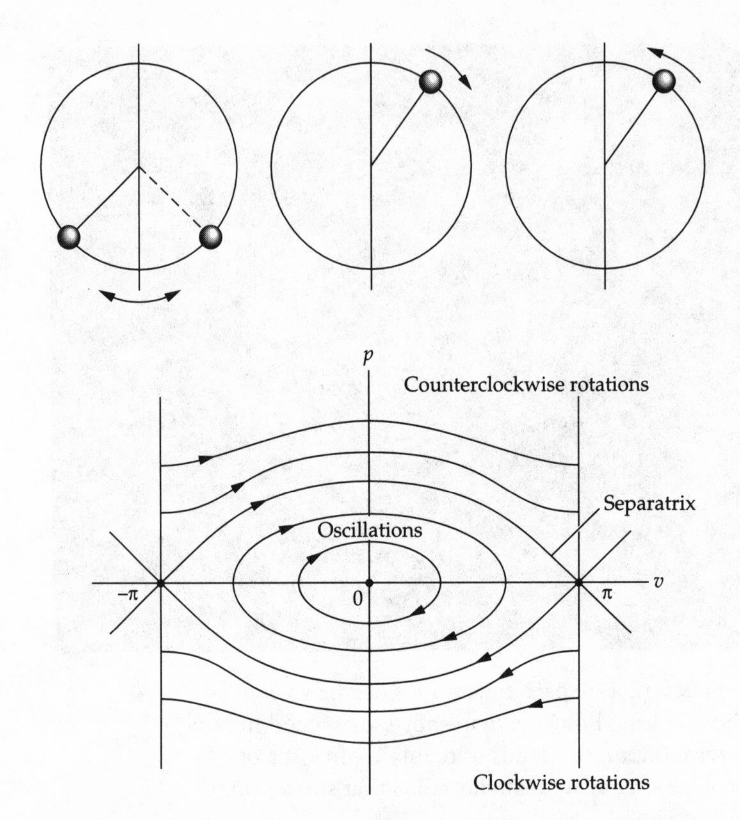

A simple pendulum can exhibit three types of motion: back-and-forth oscillations *(top left)*, clockwise rotations *(top middle)*, and counterclockwise rotations *(top right)*. In phase space, the pendulum's movements appear as closed loops. One can imagine the diagram shown *(bottom)* wrapped around a cylinder so that its left and right edges meet. The loop labeled as the separatrix corresponds to the special situation when the pendulum is precariously balanced in the unstable, upward position, moving neither clockwise nor counterclockwise.

To Poincaré, that kind of geometry implied stability, which was where he went wrong in his initial contest entry. In rethinking the matter for his revised paper, he had to confront the true meaning of these intersecting surfaces.

Richard McGehee, a mathematician at the University of Minnesota who has studied the original paper, has described Poincaré's mistake as the same kind of elementary but understandable error that many college students make today when they first encounter a similar

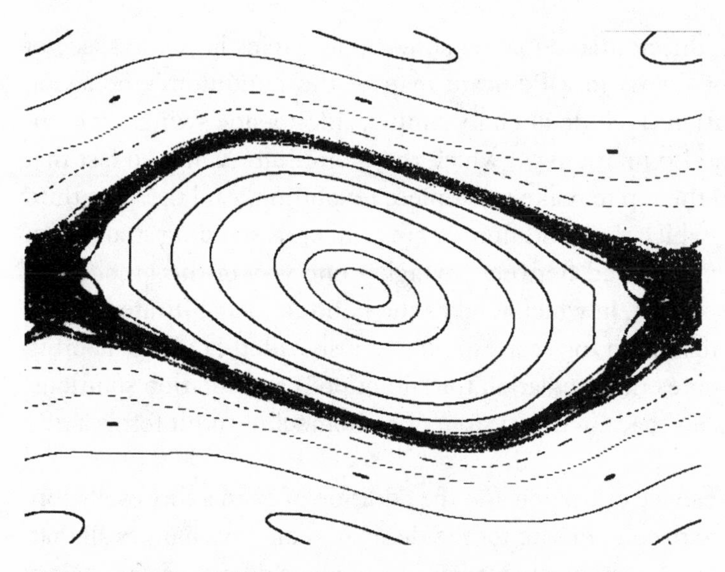

Adding a motor to drive a simple pendulum's point of attachment back and forth horizontally alters the pendulum's dynamics. In phase space, one can still see interior regions showing closed loops, which represent the pendulum's back-and-forth swings, and exterior regions, in which the pendulum swings in a full circle. But in the vicinity of the separatrix, where the pendulum nearly stalls at the top, its motion becomes chaotic and unpredictable.

situation in differential equations. But Poincaré's response was an insight of monumental import.

Because the mathematical expressions required to describe a given situation in celestial mechanics are themselves so complex, it's helpful to use the simpler example of a pendulum to understand how the question of stability is related to Poincaré's phase portraits and stroboscopic maps. Picture the kind of pendulum found in many grandfather clocks, which consists of a rigid rod pivoted at one end and carrying a weight at the other end. Gravity drives the pendulum's motion as it swings back and forth, while friction or air resistance reduces the swing's amplitude, eventually bringing the pendulum to a stop. To keep the pendulum going, an escapement mechanism periodically gives it a little kick—a small pulse of energy.

In the absence of friction and with no jolting force, a simple formula describes the pendulum's perpetual oscillations. Solutions to the corresponding differential equation show up as circles in phase space or as cycles of points in a Poincaré map of the pendulum's behavior. Those solutions include not only simple, side-to-side swings, or oscillations, but also rotations, in which the pendulum swings so fast that it goes over the top to make a complete revolution. And there's a third solution, in which the pendulum begins and ends standing straight up, precariously balanced between swinging one way or the other. This particular solution, in which it takes the pendulum an infinite amount of time to reach the top, corresponds to a so-called homoclinic orbit, which separates the oscillation (or "libration") and rotation solutions. But there's no chaos in this system. The homoclinic orbit forms a true barrier.

The presence of friction and the addition of even a tiny oscillatory driving force to compensate for the dissipated energy changes all that. The addition of small perturbations introduces a degree of uncertainty as to whether the pendulum will fall back or roll over when it nears the top of its swing. It's this sensitivity that produces very complicated motions, as the pendulum readily shifts on its dynamical knife edge from one type of motion to another. Thus, slight changes in initial conditions can produce very different results. The homoclinic barrier is broken, and orbits can now wander freely between the oscillation and rotation regimes.

Philip Holmes, an applied mathematician at Cornell University, has described this situation in these terms: "Physically, a gentle tickling of the pendulum has dramatic consequences when it is near its unstable, inverted equilibrium. . . . Each time the pendulum reaches the top of its swing, near the inverted, unstable state, the oscillation supplies a small push either to the left or right depending on the phase (time). Thus the precise time at which the bob arrives near this position is crucial, and this, in turn, is determined by the time at which it left the same position after the preceding swing. Here is the physical interpretation of sensitive dependence upon initial conditions."

What Poincaré discovered is that the fundamental equations governing the motion of three bodies incorporate similar sensitivities. Although he confined his technical discussion to the narrow context of celestial mechanics, Poincaré's reasoning reverberates throughout Newtonian mechanics. The questions asked about the moon's evolving orbit can just as easily apply to nearly any dynamical system, whether it be a pendulum driven by a motor or even, if the appropriate equations of motion weren't so hard to solve, the turbulent flow in a waterfall. Poincaré's discovery implies the astonishing notion that unpredictable, apparently lawless behavior can occur in a system ruled entirely by exact and unbreakable laws. This means that many events in the physical world are to some degree unpredictable, because it is impossible to compute the future with sufficient accuracy. Whatever the quantitative mathematical model used to predict the future, there is always an element of irreducible uncertainty at the heart of Newtonian mechanics.

Poincaré came back to these issues again and again, trying to clarify the essence and scope of the intrinsic uncertainty he had discovered. In his 1903 essay "Science and Method," Poincaré noted: "A very small cause that escapes our notice determines a considerable effect that we cannot fail to see, and then we say that the effect is due to chance. If we knew exactly the laws of nature and the situation of the universe at the initial moment, we could predict exactly the situation of that same universe at a succeeding moment. But even if it were the case that the natural laws had no longer any secret for us, we could still only know the initial situation *approximately*. If that enabled us to predict the succeeding situation with *the same approximation*, that is all we require, and we should say that the phenomenon had been predicted, that it is governed by laws. But it is not always so; it may happen that small differences in the initial conditions produce very great ones in the final phenomena. A small error in the former will produce an enormous error in the latter. Prediction becomes impossible, and we have the fortuitous phenomenon."

Poincaré's explorations of this curious twilight world of chaos and order, in which a clock's pendulum or a solar system governed by the laws of Newtonian mechanics can display such complicated dynamics, provoked a host of new questions. Even a system as simplified as Hill's special model for the moon's orbit had hidden within it the seeds of dynamical chaos. If these peculiar orbits could occur in a three-body system, then the whole solar system might be unstable. Given sufficient time, the minute effects of each planet on the others could produce the conditions necessary for an orbit to shift suddenly to a new configuration, or for the entire solar system to drift apart.

With respect to the solar system, these questions of stability and chaos are closely connected with the phenomenon of resonance. Solutions expressed in terms of infinite series fail to converge because, every so often, a term in which a large number has to be divided by a very small one unexpectedly pops up somewhere along the string. Evaluation of that term produces a huge result that can easily overwhelm whatever trend has been established by the evaluation of earlier terms. Moreover, even slight errors in the divisor add to the uncertainty by swinging the term's value over a wide range. In celestial mechanics, such terms correspond in some sense to subtle, cumulative interactions between gravitating bodies. These interactions, or resonances, reinforce one another and strongly influence the motion, much as repeated pushes at just the right frequency greatly increase a pendulum's amplitude.

In the end, Poincaré left Newton's laws of motion unchanged, but he radically altered scientists' understanding of the types of behavior they mandate. It took modern technology to provide everyone else with the means of visualizing and appreciating the importance of the world Poincaré had glimpsed in his equations.

Characteristically, Poincaré also left many matters unresolved and many conjectures unproved. His results on "chaos" encompassed most solutions, but not all. In subsequent years, mathematicians tackled a number of these issues, subduing them one by one. In 1913, a year after Poincaré's death, Finnish mathematician Karl F. Sundman actually solved the famous prize problem that had eluded Poincaré and set him on the route to dynamical chaos. Sundman found exactly what the problem asked for: a particular solution of the three-body problem that can be expressed as a convergent series. Unfortunately, this series approaches the final answer so slowly that it's useless for any practical purpose. It would take too long to work out the large number of terms needed to come close to the correct answer.

Over the next 40 years, only a handful of mathematicians continued along the course that Poincaré had set in the geometrical study of dynamical systems. Among them was George Birkhoff, who in the 1930s developed much of the theoretical basis for what is now termed chaos. As part of the British effort to develop radar during World War II, one group of researchers even stumbled upon the same sort of complicated solutions to differential equations—in this case, describing certain radio circuits—that Poincaré had found. But for the most part, few paid attention; even fewer could imagine the far-ranging implications of such erratic behavior arising out of mathematical equations used to describe physical systems.

In the 1954, Andrei N. Kolmogorov, continuing a strong Russian effort in the study of dynamical systems, revisited the three-body problem. He sketched a means of mathematically tackling the complexity associated with periodic orbits in phase space, and other mathematicians filled in the details. In 1963, Vladimir I. Arnol'd, one of Kolmogorov's students, finally managed to solve Poincaré's problem—with astonishing results. Arnol'd proved that under certain conditions, the series used to describe motions in the three-body problem did converge, but under another set of conditions they did not. Poincaré had examined only the second type when he concluded that the series typically encountered in celestial mechanics problems diverge.

Thus, depending on the initial conditions, motion in a system of three or more bodies is sometimes regular and sometimes chaotic. This regularity shows up, for example, in slow, slight changes in the eccentricity of a planetary orbit over the course of infinite time, as the entire orbit slowly rotates under the influence of perturbations and remains in a plane slightly rocking about an unchanging position. Chaos manifests itself in orbits that display abrupt jumps in eccentricity, carrying the planet or some other celestial body far afield from its usual path in the space.

Through their study of purely mathematical constructs, Poincaré and his successors unveiled a fascinating, complex realm previously hidden from view. Locked within the equations used to describe motion in the solar system, these unsuspected intricacies lay virtually dormant through two centuries of calculation—until Poincaré revealed the true breadth of Newton's laws of motion. "The true goal of celestial mechanics is not the calculation of the ephemerides [tables of the locations of planets], but rather to discover if all phenomena can be explained by Newton's law," Poincaré wrote in the introduction to his great treatise on celestial mechanics.

But did the ensuing theoretical speculations have anything to do with the real solar system of massive planets, wayward asteroids, and tumbling satellites? Did these bizarre mathematical wonders have something fundamental to say about the solar system's long-term future? After centuries of searching for regularities, the time had come to seek explicit evidence not of order but of chaos in the heavens.

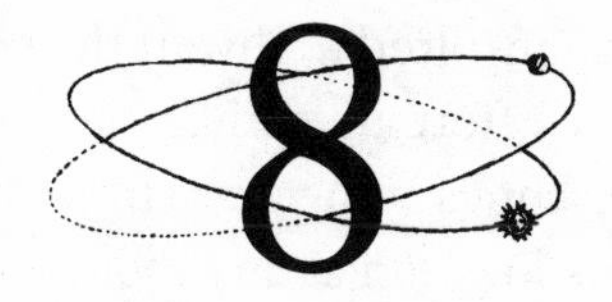

Band Gaps

•

'Tis all in pieces, all coherence gone;
All just supply and all Relation

JOHN DONNE (1573–1631),
Anatomie of the World

IN A fashion show of minor planets, the rocky fragment known as asteroid 951 Gaspra would have few distinguishing features to catch anyone's eye. A perfectly ordinary denizen of the asteroid belt, Gaspra follows an elliptical path that keeps it an average of 331 million kilometers from the sun. Originally discovered in 1916 at an observatory in the Crimea and named for the famed resort town on the Black Sea where Russian novelist Leo Tolstoy spent many years of his life, this tiny, dim speck is merely one of several thousand rocky worlds scattered between Mars and Jupiter—part of the fossil rubble left over from the creation of the solar system.

But on October 29, 1991, Gaspra had the distinction of being in the right place at the right time. On that day, the *Galileo* spacecraft, in the course of its lengthy, winding trek to Jupiter, whisked past Gaspra to catch the first close-up glimpses of an asteroid. Traveling at 8 kilometers per second, the spacecraft came as close as 1600 kilometers to this solitary, anonymous body.

Getting *Galileo* precisely to its target was a remarkable navigational feat. Uncertainties regarding its path and the limited accuracy of observations of Gaspra conspired to thwart the close encounter. But an extensive international effort to gather the necessary observations from Earth-based telescopes allowed astronomers to pinpoint their moving target so that its location at any moment was known to within a few dozen kilometers. Armed with such data, spacecraft controllers on Earth could fine-tune *Galileo*'s course to bring the asteroid into view. When the flyby finally occurred, project engineers had succeeded in bringing the spacecraft to within a celestial hair's breadth—a mere 5 kilometers—of where it was supposed to be. That precision

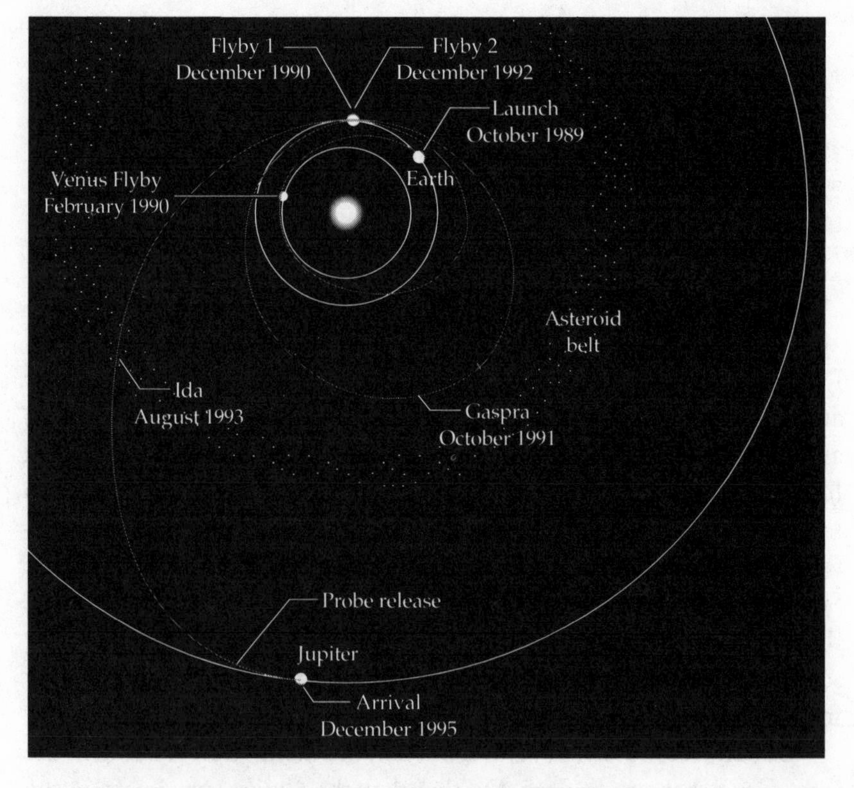

On October 29, 1991, the Jupiter-bound *Galileo* spacecraft passed near asteroid 951 Gaspra on its way through the asteroid belt. (NASA/Jet Propulsion Laboratory.)

assured them that the spacecraft's cameras would be able to snap a full portrait of Gaspra.

The first image transmitted back to Earth revealed an elongated, ravaged potato of a world measuring only 19 by 12 by 11 kilometers. Peppered with small craters, this drab, gray lump showed a surface threaded with fractures, grooves, and ridges. Its scarred landscape suggested that Gaspra had not survived intact from primordial times but was probably a fragment of a larger, much older parent body battered by collisions—a piece chipped off an old block perhaps as many as 200 million years ago.

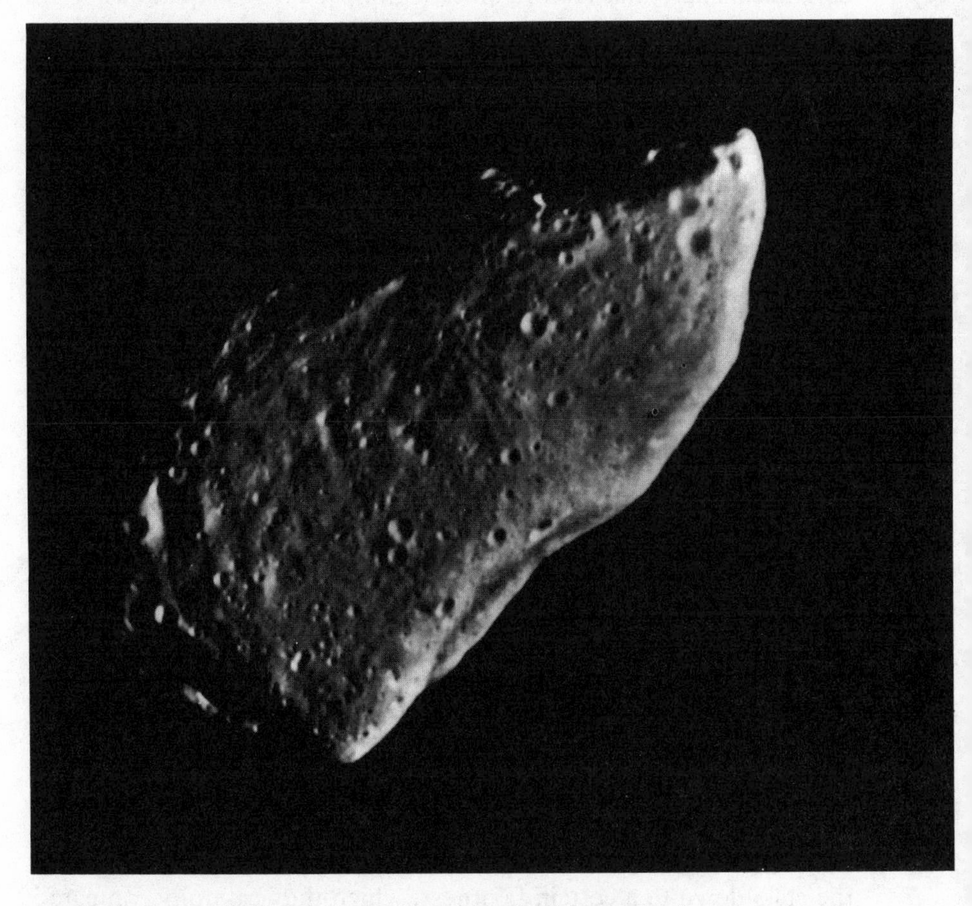

A mosaic of two images recorded when the *Galileo* spacecraft was 5300 kilometers from Gaspra, about 10 minutes before the point of closest approach, provides a detailed portrait of the asteroid. (NASA/Jet Propulsion Laboratory.)

With the help of the dramatic pictorial evidence provided by *Galileo,* together with many other observations and measurements, astronomers are beginning to piece together the puzzle of the origin of asteroids. Relics of the early solar system, these curious objects supply clues to the forces and materials that congealed and spun off the planets 4.5 billion years ago. Moreover, strongly influenced by the sun and Jupiter but themselves exerting negligibly small gravitational forces, these minor planets wander in a dynamical wonderland in which planetary perturbations can readily spread out and distort orbits that were once quite similar. What better place to look for evidence of the dynamical chaos that Henri Poincaré glimpsed in his equations?

Two hundred years ago, no asteroids had yet made their appearance in astronomical catalogs of the solar system, which included just the sun, six planets, and the mysteriously evanescent comets. But by a remarkable coincidence, the solar system itself provided a numerical clue that pointed astronomers toward the region where most asteroids are found. This hint of regularity in the solar system appealed to the human predilection for finding patterns in apparently random occurrences.

As noted in Chapter 1, Johann D. Titius saw a striking numerical pattern in the ratios of the distances of the six known planets from the sun compared with Earth's distance from the sun. But one number in this sequence had no planetary partner. There was a disturbing gap where the fifth number indicated the presence of another planet. To fill the gap and make the sequence work, Titius posited a planet between Mars and Jupiter at a distance about 2.8 times that of Earth from the sun.

William Herschel's startling discovery in 1781 of a planet beyond Saturn not far from where the numbers indicated an eighth planet would lie gave credibility to Titius's prediction and prompted a determined search for an object orbiting between Mars and Jupiter. A handful of astronomers, led by Franz Xaver von Zach, organized themselves into a self-proclaimed "celestial police" to track down the missing planet. These constables of the night sky prepared maps of all the stars down to a certain faintness, which they carefully compared with each sliver of sky they observed. They looked for objects that had not been recorded or that had evidently moved from a previously recorded position relative to the other stars.

The first sighting, however, was made not by a member of this elite task force but by Giuseppe Piazzi, a monk who in 1790 had established and equipped an observatory in Palermo on the island of Sicily. Taking advantage of a favorable climate for astronomical viewing, Piazzi had launched a lengthy project dedicated to determining precisely the astronomical coordinates of several thousand stars. Because of his facility's unique position at that time as the southernmost European observatory, he could study significantly more stars than had previously been cataloged.

On the opening day of the nineteenth century, while looking for a specific star, Piazzi noticed a considerably fainter, starlike object not included in the catalog he was checking. Observations of the mysterious object on successive nights revealed that it moved slowly against its starry backdrop, first moving backward, then reversing direction and overtaking the background stars. Unsure whether the object was a comet or a planet, Piazzi watched it regularly until February 11, when he fell ill. By the time he recovered a few days later, he was able to make only one more observation before the object advanced sufficiently close to the sun to disappear in its glare.

Piazzi had already begun to notify colleagues in other parts of Europe of his discovery, but political turmoil in Italy delayed the mails. As a result, no one else had a chance to observe the object. Only one-tenth the brightness of Uranus and already on the fringe of visibility in most telescopes of the time, this faint speck had no telltale planetary disk to make it easier to locate. To recover the object once it emerged from the sun's glare several months later, astronomers needed to know its orbit. But Piazzi's observations covered a period of just 41 days, during which time the object—badly placed for viewing—had moved through an arc of only 3 degrees across the sky. Any attempt to compute the orbit of such an inconspicuous object from this meager set of data appeared futile.

To Karl Friedrich Gauss, a 23-year-old mathematician who early in life had displayed a prodigious talent for mathematics and a remarkable facility for highly involved mental arithmetic, this problem presented an enticing challenge. Having completed his studies at the University of Göttingen, Gauss was living on a small allowance granted by his patron, Ferdinand, Duke of Brunswick. With a major mathematical work just published and little else to occupy his time during the latter part of 1801, Gauss brought his formidable powers to bear on celestial mechanics. Like a skillful mechanic, he systematically disassembled the creaky, ponderous engine that had long been used for determining approximate orbits and rebuilt it into an efficient, streamlined machine that could function reasonably reliably given even minimal data.

Assuming that Piazzi's object circumnavigated the sun on a circular course and using only three observations of its place in the sky to compute its preliminary orbit, Gauss calculated what its position would be when the time came to resume observations. In December, after three months of labor, he delivered his prediction to von Zach. Any

Karl Friedrich Gauss (1777–1855). (Smithsonian Institution.)

hope of locating this particular celestial mote after a lapse of nearly a year rested on the reliability of Gauss's innovative methods and the accuracy of his calculations. On the last night of 1801, von Zach found the object only half a degree away from where Gauss had predicted it would lie. With this dramatic sighting, the astronomer excitedly proclaimed the fugitive's recapture.

The success of Gauss's calculations vindicated his novel approach, and Piazzi's faint speck was now seen to have the nearly circular orbit of a planet rather than the characteristically elongated orbit of a comet. The calculations also confirmed that this planetlike object orbited the sun at a distance very close to that predicted by Titius for the missing planet. But it was substantially smaller than anyone had expected. In the opinion of some astronomers, such a disappointingly tiny fragment was unworthy of inclusion in the august company of the other, much grander planets.

Piazzi named his object Ceres, after the guardian goddess of Sicily. Three months later, Heinrich W. M. Olbers, a physician with a medical practice in Bremen, discovered another object similar to Ceres, which he named Pallas. Olbers, a skilled and dedicated amateur astronomer who had equipped the upper floor of his house as an observatory and regularly devoted the greater part of each night to making observations, found another such object, Vesta, in 1807. In the meantime, the asteroid Juno had also been sighted. Faced with the proliferation of these minor but impossible-to-ignore entities, William Herschel proposed the designation *asteroid* to denote their starlike appearance. Others, including Piazzi, preferred the term *planetoid* or *minor planet.*

While astronomers argued over nomenclature, Gauss continued to refine his methods, successfully calculating the orbits of Juno, Pallas, and Vesta. To Gauss, the discovery of one asteroid after another furnished new opportunities for testing the efficiency and generality of his methods. His computational triumphs brought him immediate, lasting recognition as Europe's top mathematician and a comfortable position as a professor of astronomy and director of the observatory at Göttingen, where he lived modestly for the rest of his long, productive life.

Never in a rush to see his ideas in print, whether in pure mathematics, astronomy, or physics, Gauss relentlessly reworked his results again and again until they were polished to perfection. Clarifying his thoughts step by step and eliminating everything but the essential elements, he would obliterate all traces of the path he had followed to arrive at his insights. No scaffolding ever marred the elegant mathematical structures he constructed so patiently. This austere style, which now pervades mathematics, may be one of Gauss's less happy legacies. To the uninitiated, the stark abstraction that passes for much of contemporary mathematics is virtually impenetrable. Only a select few dare to brave its dense and tightly woven thickets.

Gauss spent years refining his techniques for handling planetary and cometary orbits. Finally satisfied that they met his high standards, he noted with pride that "scarcely any trace of resemblance remains between the method in which the orbit of Ceres was first computed, and the form given in this work." Published in 1809 in a long paper called "Theoria motus corporum coelestium" ("The theory of the movement of celestial bodies"), this collection of methods still plays an important role in modern astronomical computation and celestial mechanics.

Once the initial sensation of their discovery had faded, interest in asteroids languished, and between 1807 and 1845 no more were found. But a steady trickle of sightings in subsequent years plainly indicated that these minor lights could not be ignored. Equipped with improved telescopes and increasingly complete, reliable star charts, astronomers stumbled upon more and more asteroids, and the stream of discoveries turned into a flood. By 1890, the number of known minor planets—most with similar, neighboring orbits between Mars and Jupiter—had reached 300.

To astronomers, this disconcerting plethora of minor planets was both a puzzle and an annoyance. Where had they come from? Were these the fragments of a respectably massive planet that had disintegrated some time in the past and left behind a broad trail of debris? Or was it merely leftover material that had never agglomerated into a sizable body? Whatever their number, how seriously should such obscure, apparently uninteresting objects be taken?

The introduction into astronomy of photography further escalated the rate of discovery, and by the turn of the century, the identification and tracking of asteroids had become a demanding specialty taken seriously by only a devoted few. In this astronomical backwater, tiny, faint streaks on photographic plates readily provided the evidence, but it took considerable time and attention to detail to follow up with precise position measurements, calculated orbits, and accurate predictions of future positions. In fact, characterizing orbits individually was really the only way to tell these objects apart.

It was easy to both find and lose asteroids, as the experience of Harlow Shapley and Seth B. Nicholson illustrates. In September of 1916, using the 60-inch telescope at the Mount Wilson Observatory near Los Angeles, the two astronomers were searching the night sky for compact groupings known as globular clusters. Containing millions of stars, these huge assemblages were so distant that they appeared even in the world's largest working telescope as little more than fuzzy balls. One night, Shapley and Nicholson by chance detected what turned out to be a very faint asteroid. By this time, astronomers were starting to consider these minor planets nuisances rather than phenomena of real astronomical interest. More as a lark than for any serious purpose, Shapley and Nicholson decided to observe the asteroid as part of their nightly routine, periodically checking its position as it slowly drifted across the sky relative to the stars. Shapley named the object, designated number 878 on the master list of known asteroids, after his one-year-old daughter Mildred.

By the time the asteroid disappeared from view six weeks later, the astronomers had accumulated sufficient data to enable them to compute a preliminary, approximate orbit, which indicated that Mildred travels along a rather elongated course just outside the orbit of Mars. So far as Shapley was concerned, this ended the matter.

Despite several efforts by other astronomers in subsequent years to find asteroid Mildred, it remained incognito for decades (and Mildred Shapley was teased by her four brothers and her astronomer friends about being a "lost" woman). Seventy-five years after its initial discovery, however, on May 25, 1991, the International Astronomical Union issued a notice announcing that asteroid Mildred had finally been found—the result of an impressive piece of astronomical detective work.

Gareth V. Williams, associate director of the Minor Planet Center at the Harvard-Smithsonian Observatory in Cambridge, Massachusetts, had received the coordinates of several asteroids recorded on a photographic plate exposed on April 10 at the European Southern Observatory's 1-meter telescope in Chile. Williams's calculations led

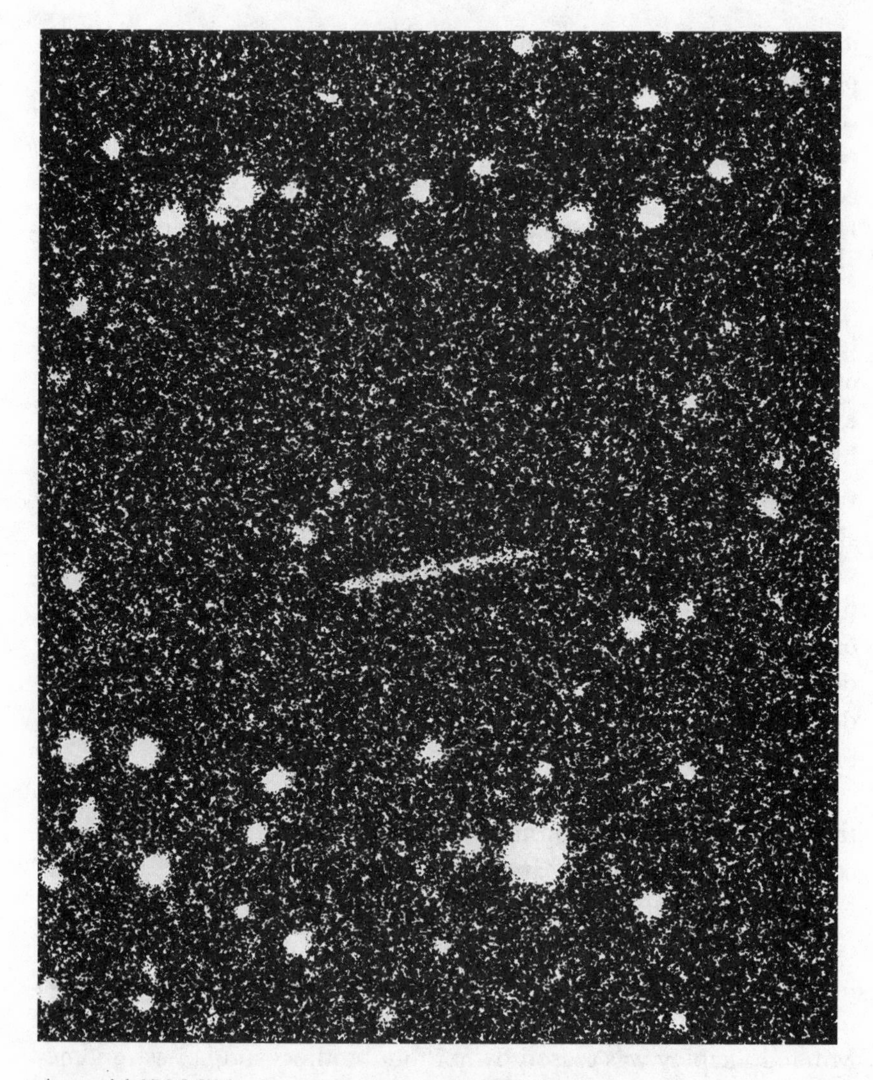

Asteroid 878 Mildred appears as a streak in this enlargement of a grainy photograph taken on June 11, 1977. Because the exposure required 60 minutes, Mildred left a trail as it moved relative to the background stars. (European Southern Observatory.)

him to believe that the motion of one of these asteroids fit an extrapolation of the 1916 measurements of Mildred. Like a number of other astronomers, Williams had been on the lookout for Mildred and had developed a sense of where it was likely to be. With another intuitive leap and extensive work at the computer, Williams discovered that an asteroid observed once in 1985 also seemed compatible with the general track followed by Mildred. Once they knew what they were looking for, other astronomers quickly found additional images of the asteroid on photographic plates obtained in 1977 and 1984. That was enough to establish that Mildred was back.

The news, of course, delighted Mildred Shapley Matthews. Now a research assistant and editor at the Lunar and Planetary Laboratory at the University of Arizona in Tucson, she remarked, "I promised to hang in there until it was found again."

Losing asteroids was in fact a common occurrence. In 1931 alone, for example, asteroid hunters found 398 new objects, but only 159 were characterized well enough to remain under surveillance for an appreciable period. The rest, after a brief moment in the spotlight, simply blended back into the anonymity of the asteroid belt. The huge task of keeping track of these "vermin in the sky" seemed hardly worth the trouble. The numbers alone were staggering and daunting.

Dirk Brouwer, one of the premier practitioners of celestial mechanics in this century, commented in 1935, "A question that is being asked with more emphasis at the present time than ever before is: What are the astronomers going to do with those hundreds of minor planets to which tens of new ones are being added every year? All these orbits require constant care. Even the calculation of accurate ephemerides for over a thousand objects from elliptic orbits is not a light task. The calculation of the theoretical positions with an accuracy corresponding to the best modern observations requires that the attractions of the principal planets be taken into account. Doing this for all the known planets represents an amount of labor that, at the present time, seems prohibitive." Even with today's digital computers, the need for accurate predictions of asteroid positions—used as reference points for spacecraft instruments and other purposes—and the rapid rate of discovery of new asteroids make Brouwer's comment still ring true.

Astronomers have now identified and characterized more than 5000 asteroid orbits, and they estimate that the main asteroid belt contains as many as a million chunks of rock with diameters of a kilometer or more. Given so many asteroids, it's easy to imagine the asteroid belt as a roller derby of crashing bodies. In reality, however, the volume of space through which these objects pass as they whirl around the sun is so vast that any one asteroid is typically several

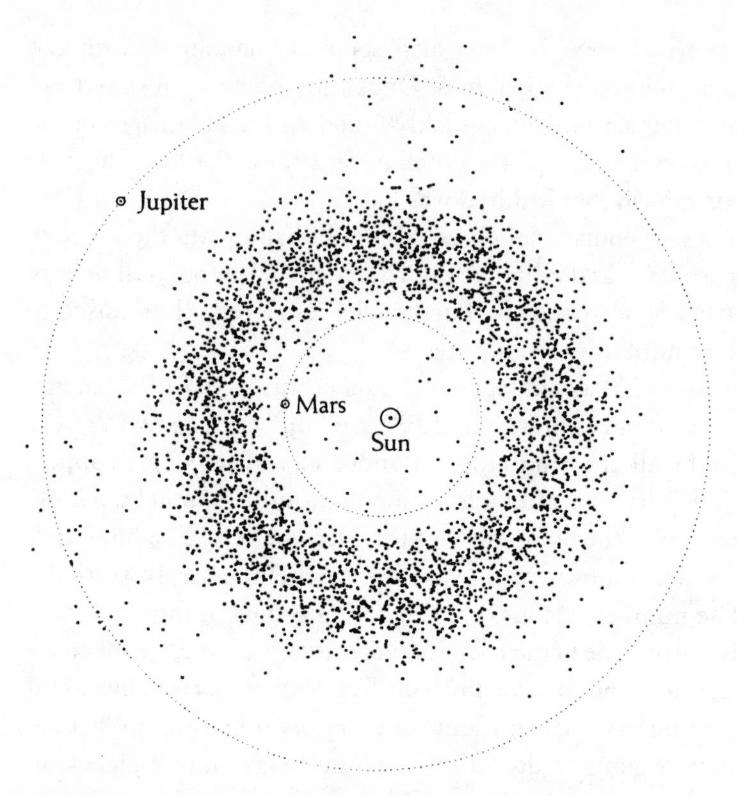

A "snapshot" of the inner solar system capturing the locations of the sun, Mars, Jupiter, and approximately 5000 asteroids at one moment in time. Most asteroids orbit the sun in a belt 1.5 astronomical units wide between the orbits of Mars and Jupiter. (Courtesy of Dr. S. Ferraz-Mello, University of São Paulo.)

million kilometers away from its nearest neighbor. Close encounters and true collisions are relatively rare.

Nonetheless, strange things can happen in the asteroid belt. Repeatedly pulled one way and then another by the competing gravitational tugs of the sun, Jupiter, and (to a lesser extent) other planets, asteroids wander a rippled path in their repeated sweeps around the sun. On such courses, subtle influences over long periods could conceivably accumulate to cause radical changes in orbits.

The painstaking observations and insightful analyses of Daniel Kirkwood provided the first clues that the asteroid belt harbors surprising dynamics. An enthusiastic and inspiring teacher of mathematics and astronomy, Kirkwood spent much of his spare time studying the roles of the solar system's lesser members—comets, minor planets, and meteorites. While at Indiana University in 1857, he noted puzzling fluctuations in the number of asteroids at various distances from the sun.

Because asteroids generally travel along elliptical paths, their distances from the sun vary over a certain range. This variation smears any gaps that might be visible on a plot of actual asteroid positions at any given moment. Instead of looking at position, Kirkwood listed the value of each asteroid's semimajor axis, which represents the distance from the center of its elliptical orbit to its furthermost point from the sun. To his surprise, the columns of figures revealed a paucity of asteroids at certain values of the semimajor axis.

Kirkwood noted that of the 50 asteroids known at the time, surprisingly few orbit the sun at distances closely corresponding to periods of revolution one-third, two-fifths, and one-half that of Jupiter. These

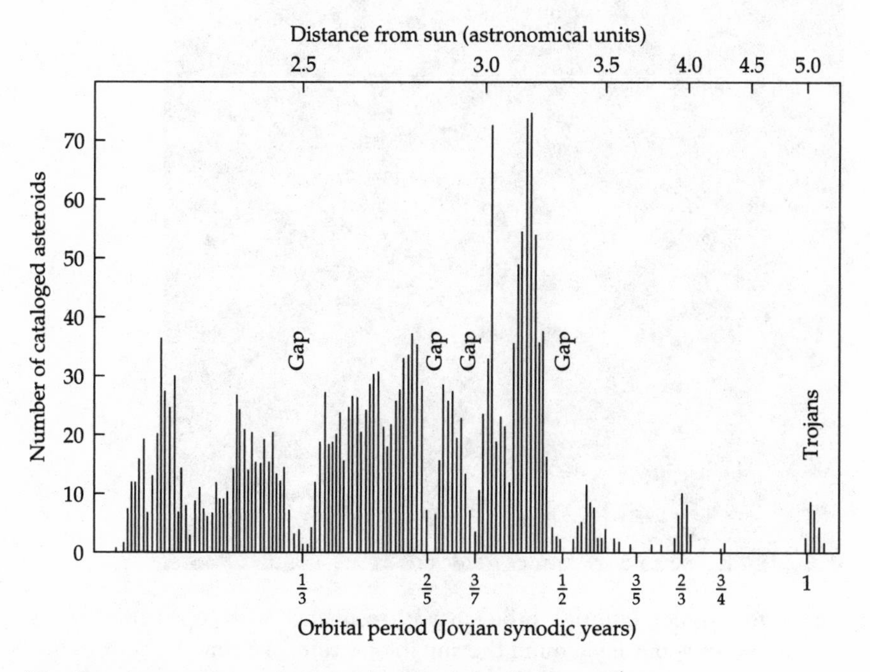

Plotting the numbers of asteroids at various distances from the sun reveals that very few have orbits whose periods correspond to such simple fractions as 1/3, 2/5, 3/7, and 1/2 of Jupiter's orbital period.

gaps are located at precisely those distances from the sun at which the ratio of an asteroid's orbital period to that of Jupiter would be one of small whole numbers. For example, if an object happened to follow an orbit about 2.5 times Earth's distance from the sun, it would complete three circuits for every one that Jupiter completes. Any objects that orbit the sun at roughly this distance would lie in a so-called 3:1 resonance.

Such resonances are important because these simple whole-number ratios mean that asteroid and planet come close to each other at almost the same place at regular intervals. Over time, these coinci-

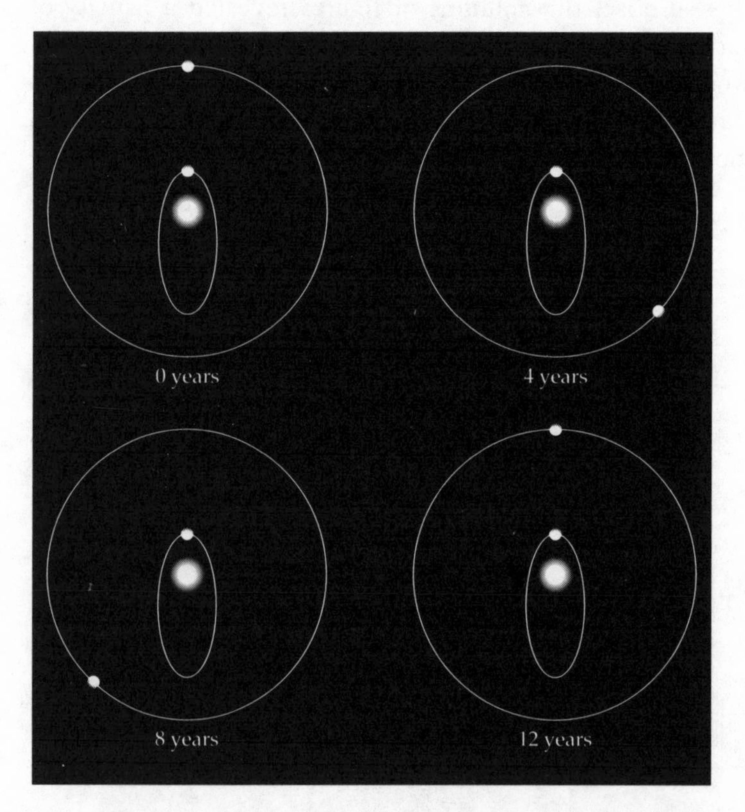

An asteroid (inner, elliptical orbit) in a 3:1 resonance with Jupiter completes a journey around the sun in one-third the time it takes Jupiter (outer, roughly circular orbit). In this hypothetical example, the asteroid and Jupiter are closest together every 12 years. Such repeated tugs from the planet can fling the asteroid into a new, potentially Mars-crossing orbit.

dences accumulate into perceptible gravitational effects that could displace the asteroid sufficiently to change the eccentricity of its orbit and shift its observed position in the sky.

Subsequent studies by Kirkwood and others revealed additional asteroid deficits at resonances where the ratio of Jupiter's period of revolution to an asteroid's would be 7:3 and 9:4. Curiously, certain ratios, such as 1:1 (marked by the presence of clumps of asteroids, known as the Trojans, which follow the same orbit as Jupiter but stay well out of its way) and 3:2 (the Hildas), point to an overabundance rather than a shortage of asteroids at the corresponding distances.

Kirkwood suggested that repeated tugs from Jupiter somehow dislodge asteroids from the vicinity of a resonance. Perhaps Jupiter's motion perturbs any bodies in such orbits sufficiently to cause instabilities that sweep them away to distances at which resonance no longer occurs. But this explanation didn't satisfactorily account for deficits at some resonances and clumping at others, and it provided no picture of precisely how gravity could effect such a potentially dramatic change in an asteroid's orbit.

One way to get a handle on the stability of typical asteroid orbits is to focus on how their orbital shapes evolve over eons. Early efforts in this direction relied on a kind of averaging of the mathematical expressions, derived from perturbation theory, that approximately describe an orbit. This technique is similar to keeping a camera's shutter open for a long time to photograph a seascape. In the resulting photograph, the constant rise and fall of the waves averages out to a glassy smooth surface. However, depending on the duration of the exposure, long-term trends that haven't been smoothed out remain visible as definite shifts in the horizon's position. The trick in using this technique is to pick the appropriate averaging period in order to highlight particular long-term variations. In celestial mechanics, the absence of trends away from the average would suggest long-term stability.

Although a number of astronomers tried to apply the principle of averaging to the asteroid conundrum, Henri Poincaré's brilliant analysis of the flaws inherent in perturbation theory (outlined in the previous chapter) called such procedures into question. Because small differences in initial conditions could lead to great disparities at future times, it was impossible to predict with any degree of assurance the changes an orbit might undergo at some point in the distant future. Nonetheless, the deviations are quite small over short periods spanning just a few hundred years, and Poincaré himself used the averaging technique in an attempt to determine how the 2:1 resonance affects asteroid orbits. But he had little success.

With the apparent failure of approaches based on orbital mechanics to explain the mysterious gaps in the asteroid belt, a growing number of astronomers adopted a position that conveniently removed the problem from dynamical theory. If these gaps hadn't evolved from the way asteroids behave under the influence of gravity, they argued, perhaps the gaps were there to start with when the solar system formed, or perhaps they were linked in some way to the processes that created the solar system itself.

In the 1970s, however, mathematical astronomers and planetary scientists began to take a fresh look at the evolution of orbits. They had a new tool—the computer—that permitted them to use the fundamental equations of motion to compute orbits farther into the future than Poincaré and his immediate successors had been able to manage. Furthermore, they were awakening to the notion that purely mathematical tools, developed largely by mathematicians to pursue a variety of dynamical problems, could be relevant to physical systems—perhaps even to the solar system. This mathematical realm, which included but was far from limited to what we now call chaos, suggested a novel, provocative framework within which to ponder celestial mechanics.

Although Poincaré had articulated many of the basic ideas underlying the notion that deterministic systems could sometimes display unpredictable behavior, few scientists had bothered to struggle through the technicalities of their predecessor's massive works on celestial mechanics to recover those ideas. Practically no one outside of a small community of mathematicians, who had followed in Poincaré's footsteps to complete and extend key parts of his work, paid much attention to these intriguing developments or realized their profound implications for the description of physical systems.

Now it was the astronomers' turn to wander not in "real" space but in the mathematicians' phase space, within which they could visualize the changes in an object's position and velocity as a winding trajectory. What became clear in these early computer studies is that the phase space in which planetary and asteroid positions and velocities are plotted could generally be divided into regions of regular, predictable behavior and regions of chaotic, unpredictable behavior. For some combinations of starting points and initial velocities, the future was clear; for others, it remained cloudy. In this abstract realm, resonances often appeared as tiny islands of apparent stability surrounded by zones of chaos.

The trouble was that it took a tremendous amount of computer power to integrate the differential equations required to calculate asteroid orbits and to map the chaotic regions that were undoubtedly present. Researchers could reach perhaps 10,000 years into the future, but they had no assurance that this was long enough to capture whatever dramatic shifts might occur—if they occurred—in the shapes of orbits. There was no definitive answer, for example, to the key question of whether asteroids could leave certain orbits to create the gaps evident in Kirkwood's survey. But there were enough clues to suggest the importance of clarifying, even in a qualitative way, the nature of asteroid motion near resonances before any further hypotheses concerning the origin of the Kirkwood gaps were seriously entertained.

Only calculation could provide the answers. Was there a way of pushing the numerical integration of asteroid orbits 100,000 instead of merely 10,000 years into the future? Could one cleverly and efficiently smooth the numerical results to remove most of the rapidly varying contributions while retaining components due to resonance effects and long-term trends? Was there a computational shortcut to the future?

Resolution of some of these problems came in the trail-blazing work of a graduate student at Caltech in the late 1970s. At that time, the gospel of dynamical chaos was slowly beginning to spread from a small band of enthusiasts to the larger scientific community. Among those who caught the chaos bug was Jack Wisdom. Deeply immersed in celestial mechanics and intrigued by the notion of irregular behavior in a deterministic system, he readily succumbed to its mysterious allure.

It was Wisdom's thesis adviser, Peter Goldreich, who first suggested to him the potential applications of chaotic dynamics to problems in celestial mechanics. Wisdom recalls being particularly struck by Goldreich's description of the Kirkwood gap problem as a promising candidate for such an approach—*if* one could figure out a way of computing the essential characteristics of asteroid orbits far more rapidly than available methods and computer equipment allowed.

That was a formidable obstacle, but Wisdom persevered. Weaving ideas from several sources into a new, unique fabric, he step by step developed the computational shortcut necessary for studying the evolution of asteroid orbits over the course of millennia. In effect, he found a simpler, readily computable mathematical substitute for the differential equations of celestial mechanics. Then, like Poincaré, he focused not on continuous flows from one point in phase space to the next but on stroboscopic snapshots of the system's behavior, which collectively captured the crucial features without bringing in irrelevant details (see Chapter 7). Assuming that his surrogate method truly represented the behavior of the underlying differential equations, Wisdom could accomplish in seconds of computer time what had taken earlier investigators days to achieve.

The initial focus of Wisdom's studies was the behavior of asteroid orbits in the neighborhood of the famous 3:1 resonance, where Kirkwood had found one of the more striking gaps. To track asteroid movements, he populated the region of the 3:1 resonance with about 300 fictitious, massless objects and calculated how their orbits evolved over two million years. Each calculation of an asteroid orbit with a slightly different starting position in the region of this resonance provided an additional point in Wisdom's plot of how readily various orbits could change their fundamental shape, or eccentricity. He discovered that, in some cases, orbits could evolve without radically changing their eccentricities. These orbits apparently lay in a stable portion of phase space. For other starting points and velocities, orbits sometimes shifted abruptly from one eccentricity to another. Such orbits lay in a chaotic region. From this intermingling of order and chaos, Wisdom could begin to gauge the probability of an asteroid's expulsion from its initial orbit.

Using these calculations, Wisdom prepared a detailed plot showing the extent of the chaotic regions corresponding to unstable orbits in the vicinity of the 3:1 resonance. His calculations showed that Jupiter induces a chaotic zone, defined by a range of initial particle velocities and positions, in which two particles with almost identical initial positions and velocities can end up in very different orbits. The future

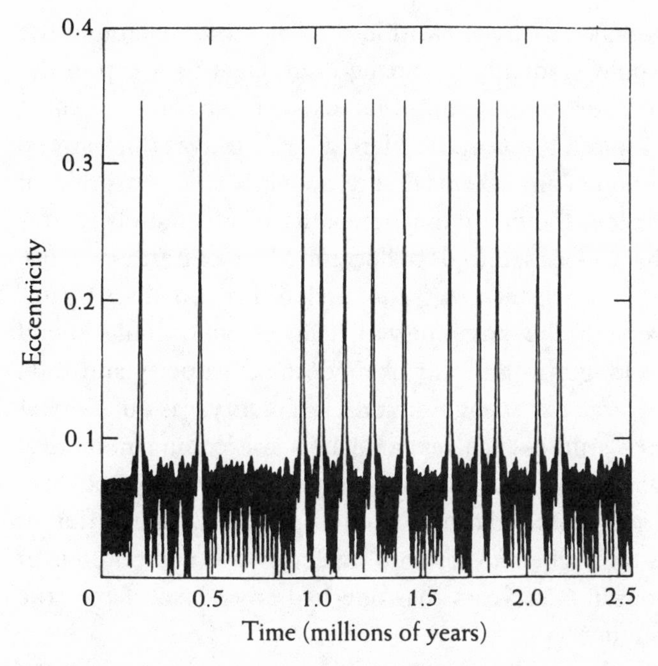

Spikes in the plot of the eccentricity of a hypothetical asteroid's exceptionally chaotic orbit near a 3:1 resonance with Jupiter, computed over millions of years, correspond to sudden large changes in the shape of its elliptical orbit. (Courtesy of Jack Wisdom.)

course of any fragments, created in occasional collisions between chunks of rock, that happen to end up in the appropriate place with the right velocity becomes unpredictable over long time periods. That doesn't mean that these asteroids will necessarily change their orbits, but it leaves open the possibility of such a drastic event.

Wisdom's numerical experiments with his fictitious, massless asteroids showed that an asteroid in the chaotic zone can spend as many as a million years exhibiting a low-eccentricity, near-circular orbit. But that languid behavior can be unexpectedly interrupted by a few irregularly timed jumps, during which the orbit becomes highly eccentric. Such behavior can have dire consequences. Near-circular orbits can suddenly stretch after a few hundred thousand years of placid behavior, becoming so elliptical that they cross the paths of Mars and Earth. Over the years, these temporarily displaced asteroids get swept up by the planets, crashing into Earth or Mars or getting hurled into new orbits.

Thus, orbits that initially fall within the chaotic zone around the 3:1 resonance would gradually be cleaned out, creating a gap in the asteroid belt. In other words, the 3:1 gap exists not because of some direct action by Jupiter, but because Mars and Earth, over time, sweep up a large proportion of any objects that temporarily shift into orbits of sufficiently high eccentricity to cross their paths. Jupiter merely creates the resonance that causes asteroids to become Mars- or Earth-crossers. Wisdom's model plausibly accounts for much of Kirkwood's 3:1 gap.

The discovery of the possibility that an asteroid could spend 100,000 years or longer in an orbit of modest eccentricity and then suddenly make a large excursion from near circularity was quite unexpected. Wisdom's studies also revealed how misleading numerical integrations that spanned just 10,000 years could be. Viewed over much longer periods, asteroid orbits could display amazingly different behavior. It was only when an asteroid trajectory was computed over hundreds of thousands of years that one had any sense of the true nature of its motion.

Wisdom also calculated that one out of five of the objects from the chaotic zone around the 3:1 resonance could end up on an Earth-crossing orbit within a period of half a million years. That finding suggested a solution to the long-standing puzzle of the origin of meteorites. Although many suspected that the most common type of meteorite, known as a chondrite, comes from the asteroid belt, no one could adequately explain how any of these rock fragments could get into orbits that cross Earth's path without invoking elaborate, unlikely scenarios. The beauty of Wisdom's mechanism was that it didn't involve any complicated procedures requiring more than one close approach of an asteroid to any planet.

When Wisdom first rather cautiously presented this startling idea in his 1981 thesis and in a paper published the following year in *The Astronomical Journal*, most astronomers were skeptical. Many thought that such unusual behavior must be an artifact of the mathematical method used to make the predictions rather than a sign of real events. They suspected that his mathematical sleight of hand didn't actually simulate the true motions of asteroids. Instead, it simply provided a means of computing the extent and nature of the chaotic zones around a resonance, without specifying where an asteroid would be at any given time. The critics argued that there was no direct correspondence between the trajectories Wisdom had mapped and those resulting directly from integration of the differential equations.

Indeed, evidence of the presence of chaos actually exacerbated the problem of computing chaotic trajectories accurately. Consider a billiard table studded with large, closely spaced cylindrical bumpers. Two balls with slightly different starting points will generally end up following quite different paths as they ricochet through this array. Predicting where a given ball will go is a delicate matter if there is any uncertainty concerning its initial position. After a few bounces, the errors will pile up so much that calculations of the ball's future position can do no better than suggest a range of likely positions, even though the ball itself follows a particular, well-defined course.

Moreover, if a small change in starting point leads to a strikingly different result, then the slight differences caused by rounding off numbers in a computation will have a similar effect and may also influence the final result. Consider a computer working with numbers to an accuracy of 14 decimal places. Computer experiments show that two neighboring trajectories starting at points differing only in the last decimal place will look totally unrelated after a few dozen steps. Just as the errors mount with every bounce of the billiard ball, so do they escalate with every computational step that requires rounding off a number.

Thus, a researcher must somehow untangle the consequences of rounding errors in mathematical calculations, of uncertainties caused by limitations in the measurements of physical systems, and of intrinsic characteristics of the physical system itself. The question of predictability is significant because the iteration, or repeated evaluation, of appropriate mathematical expressions is a standard method of finding approximate solutions of equations used to describe systems in celestial mechanics. How good can those predictions be when the initial conditions are generally known only to one or two decimal places and when the answers generated by a computer may be intrinsically uncertain?

Consequently, long numerical integrations of chaotic trajectories in celestial mechanics are rarely reversible. It isn't possible to compute an asteroid's orbit 200,000 years or more into the future and then work backward to exactly the same position and velocity at which the exercise began. If the system neither gains nor loses energy—and the notion of conservation of energy lies at the foundation of celestial mechanics—then the computations ought to be completely reversible. The cumulative errors caused by rounding off numbers, combined with the rapid divergence of neighboring trajectories characteristic of chaos, rule otherwise.

In such situations, mathematical astronomers counting on the computer to produce useful results can't completely and rigorously justify their methods. Whereas traditionalists, steeped in the ways of Laplace and Lagrange, would shudder at the thought of using a nonreversible trajectory, a modern dynamicist would say that it depends on what you want to get out of the computation. The new methods can't pinpoint an asteroid's precise location and velocity a million years into the future or the past, but they can demonstrate the likelihood of sudden shifts in an orbit's eccentricity. Furthermore, even though a computed trajectory may be inexact, it still behaves qualitatively like the true trajectory, hopping around phase space and filling nearly the same region—so long as certain precautions are taken when computing the trajectory. Sometimes this requires that numerical results be left in the form of long strings of digits in order to keep round-off error from striking a fatal blow.

Jack Wisdom noted, "Fortunately, this does not (just) mean that the hands are clasped in a fervent prayer to the god of dynamics that the results are correct. Rather one must go as far as possible to bring the phenomenon into the broader context of dynamical systems and show that the behavior is in fact not extraordinary, but just like that found in other dynamical systems." To get additional numerical results to support his claims, Wisdom scrounged enough computer time to perform the necessary calculations using the full, unaveraged differential equations—a technique more familiar and acceptable to the astronomical community. He got essentially the same results as he had using his shortcut.

Wisdom concluded one of his papers with these words: "Confronted with the evidence . . . , it is impossible to discount the importance of chaotic behavior in the formation of the gap [in the asteroid belt]. No extra hypotheses are needed beyond the dynamics of the three-dimensional elliptic restricted three-body problem and the presence of Mars to obtain the 3/1 Kirkwood gap's precise size and shape."

George Wetherill was one astronomer who took Wisdom's ideas seriously even before the extra calculations were performed. He used Wisdom's model to calculate what deviant orbits near the 3:1 resonance would look like when the asteroids approached or struck Earth. The results closely matched Earth-based observations of meteorite tracks in the sky, suggesting that Wisdom's mechanism works. The zone surrounding the 3:1 resonance conceivably serves as the source of at least some of the debris that flashes through Earth's atmosphere as meteorites.

Nonetheless, Wisdom's hypothesis didn't completely settle the matter. Although dynamical evidence points to a region near the 3:1 gap as the source of some Earth-crossing asteroids, the composition of chondritic meteorite remnants found on Earth, as determined by laboratory tests, doesn't exactly match that of asteroids presently orbiting near the gap, as deduced from studies of light reflected off representative bodies.

There are other problems as well. The 2:1 and 3:2 resonances are far more complicated dynamically than the 3:1 resonance, which lies relatively far from other important resonances. Nonetheless, the equations used to describe asteroid orbits near these two resonances are practically identical, and the solutions of both sets of equations are qualitatively similar. Yet the 2:1 resonance marks a gap in the asteroid belt, whereas the 3:2 resonance shows an overabundance of asteroids—a group known as the Hilda asteroids.

Why is there a paucity at one resonance and a surplus at the other? The true origin of this difference in population remains elusive, although recent computations have pointed to qualitative differences in the phase spaces surrounding the two resonances. In particular, the 3:2 zone appears largely devoid of chaotic behavior, which may account for the presence, if not the increased number, of asteroids. The Hildas may simply have settled into a dynamical niche safe from chaos. Curiously, several simplified computational models of the 3:2 resonance predict even more asteroids than there really are.

It also isn't clear why there should be a gap at the 2:1 resonance, because chaotic orbits by themselves aren't enough to ensure one. In the 3:1 case, the shift to highly eccentric orbits brings the asteroids across the orbits of Mars and Earth, and the planets sweep them away.

Asteroids orbiting near the 2:1 resonance are much farther from the sun and have a much smaller chance of ending up in elongated orbits reaching all the way to the orbits of Mars and Earth. What causes the demise of the 2:1 asteroids?

There is also no explanation for another puzzling feature of the distribution of asteroids; namely, the marked decline in their number beyond the 2:1 resonance. The most obvious explanation is that Jupiter's perturbing influence is too strong, and any orbits in this region would be unstable. However, this account doesn't hold up. Early numerical studies found that orbits in the region between the 2:1 and 3:2 resonances are not especially chaotic or unstable. On the other hand, these studies may have covered too short a period in the evolution of the orbits to detect instability. Each strip of the asteroid belt could very well have unique dynamical characteristics that depend on subtle differences in the gravitational forces acting on the resident bodies.

Moreover, individual asteroids can convey perplexing messages. In 1992, Andrea Milani and Anna M. Nobili, researchers at the University of Pisa, took a close numerical look at the orbit of asteroid 522 Helga, which orbits in a sparsely populated zone of the outer asteroid belt. They showed that this particular object follows a "chaotic" orbit, in that its position cannot be precisely predicted more than 10,000 years in advance. Yet the asteroid, which follows an elliptical path relatively close to Jupiter's orbit, appears to be confined to a certain region of the asteroid belt. Such curious behavior leads to a peculiar terminology and to the apparently paradoxical concept of "stable chaos."

In commenting on the research, astronomer Carl D. Murray described this strangely dissonant behavior: "Mention of chaos in the motion of objects in the Solar System conjures up images of colliding planets, impacting asteroids and a variety of catastrophic events. However . . . although gravitational perturbations can lead to chaotic orbits, the nature of the chaos can be quite subtle." He went on to remark that the apparent paradox of stable chaos lies more in the imprecision with which we use the two words in everyday life than in any real contradiction in dynamical theory. Given the many different ways of defining stable, it's quite easy to imagine an orbit that displays erratic deviations from a regular track, all of which take place within strict limits. For example, a ball rattling around a roulette wheel takes essentially a chaotic trajectory yet usually remains confined to the wheel.

In the case of asteroid Helga, the fact that the orbital periods of the asteroid and Jupiter lie roughly in a 7:12 ratio produces a resonance that somehow shelters the asteroid from disruptive, close approaches to the planet. Extending their studies of Helga to more general dynamical issues, Milani and Nobili went further by arguing that stable chaos may be a rather common feature of solar system dynamics.

Much of the current research on asteroid orbits is concerned with explaining why the solar system has the arrangement it does. By carefully tracing out the boundaries of chaotic zones, researchers can better understand the dynamics of asteroids, eventually accumulating sufficient information to examine their primitive origins.

Such studies may also have some practical value because of the growing interest in searching for near-Earth objects. In the last few decades, astronomers have identified a surprisingly large number of objects moving in orbits that bring them quite close to Earth. Although the likelihood of a collision with a given object is minuscule, it remains a remote possibility, and Earth's geologic record provides strong evidence of past catastrophes resulting from such crash landings. Ancient craters and other scars attest to the tremendous energy that such collisions involve. Scientists have recently come to the startling realization that a body about 10 kilometers across may have crashed into Earth about 65 million years ago, triggering an environmental catastrophe that initiated the mass extinction of dinosaurs and many other species. One dramatic event occurred on June 30, 1908, near the Tunguska River in Siberia. A rapidly moving object, possibly a stony asteroid about 30 meters across, exploded at a height of roughly 9 kilometers, creating no impact crater but flattening about 1000 square kilometers of forest. Cataclysms on this scale occur perhaps once every few centuries.

Now that astronomers are starting to pay closer attention to such objects, they have noted several near misses in recent years. On January 18, 1991, for example, they detected a small object, designated 1991BA, that came within 170,000 kilometers of Earth, less than half the distance to the moon. Only 5 to 10 meters across, it was the smallest

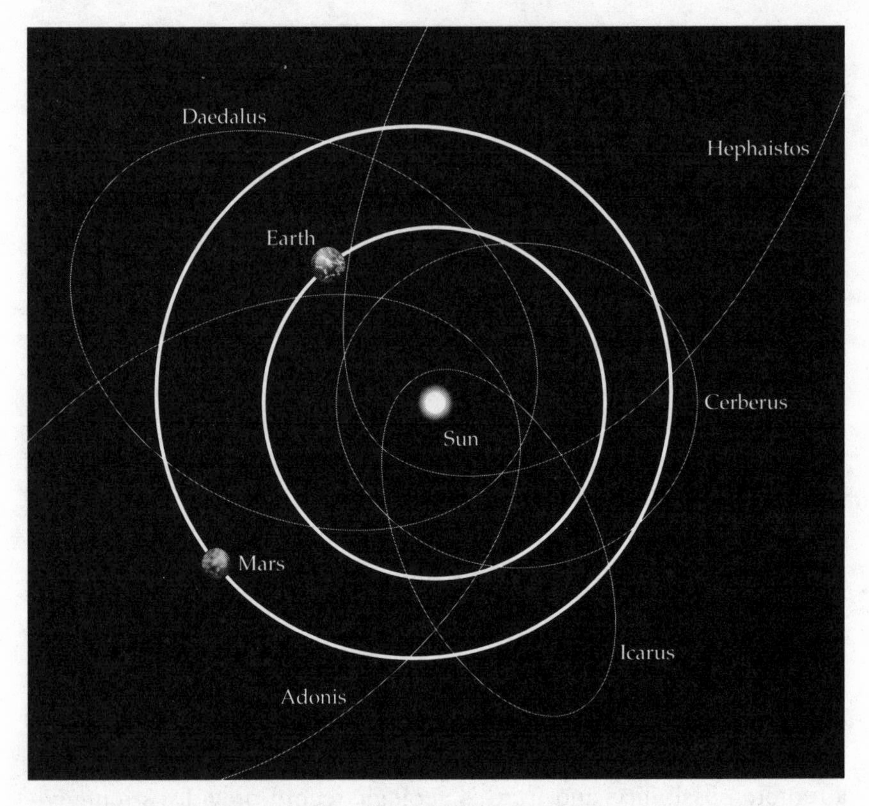

Apollo asteroids, like the five indicated here, have orbits that cross Earth's path. More than 100 have now been tracked. (Lucy McFadden, California Space Institute.)

asteroid ever observed, but if it had hit Earth, it would still have packed quite a wallop.

Why such near-Earth asteroids exist at all remains a puzzle. Any asteroids with orbits within the inner solar system are likely to collide with Mercury, Venus, Earth, or Mars within a hundred million years or so. Because this time span is much shorter than the 4.5-billion-year age of the solar system, any objects present at its beginning would long ago have been swept up by the planets. Astronomers suspect that some process is continually throwing new objects into planet-crossing orbits to make up for the ones destroyed in collisions. This resupply route probably originates somewhere in the asteroid belt, with Jupiter's gravitational influence serving as a kind of catapult.

On the other hand, some of these bodies may be the dead hulks of extinct comets—dirty snowballs of ice and dust typically a few kilometers across—disguised as ordinary, rocky asteroids. Their coats of dust may be so thick that they turn off the evaporation process that creates a comet's distinctive fuzzy coma and tail. Astronomers hope to obtain an idea of the composition of at least some of these objects by carefully measuring the spectrum of sunlight glinting from their surfaces. Each material present removes a tiny bit of light at characteristic wavelengths to create telltale gaps in the spectrum, which serves as a kind of fingerprint.

So far, astronomers have identified more than 130 near-Earth objects, none of which is expected to hit Earth within the next 200 years. But to achieve greater certainty, astronomers see a critical need to monitor the motions of known near-Earth objects more closely and to find other examples with which to build up a more complete picture of their distribution. As telescope technology improves, astronomers are likely to report even more near misses in the future.

So the dynamics of asteroids, especially near resonances, remains puzzling. Clearly, all resonances have features that somehow distinguish them from others, but unraveling their true nature is a formidable undertaking. Although researchers have in general found reasonably good qualitative agreement between the locations of gaps and chaotic zones, quantitative comparisons often don't show perfect agreement. That leaves open the possibility that the processes responsible for creating the solar system may have played a role in producing the structure and special characteristics of the asteroid belt.

At the same time, it's useful to note that the regions of the asteroid belt where the motion is unpredictable are relatively small. Most asteroid orbits appear stable. Only those near resonances show irregular behavior, and even that isn't universal, as the case of the surplus asteroids observed at certain resonances and the stable chaos detected by Nobili, Milani, and others show. Nonetheless, without a dynamics that includes chaotic motion, there would be no way for gravity to deliver meteorites and comets from the asteroid belt to impacts on Earth and elsewhere and to send them on long, looping voyages around the sun, or for planets to capture itinerant bodies and retain them as satellites.

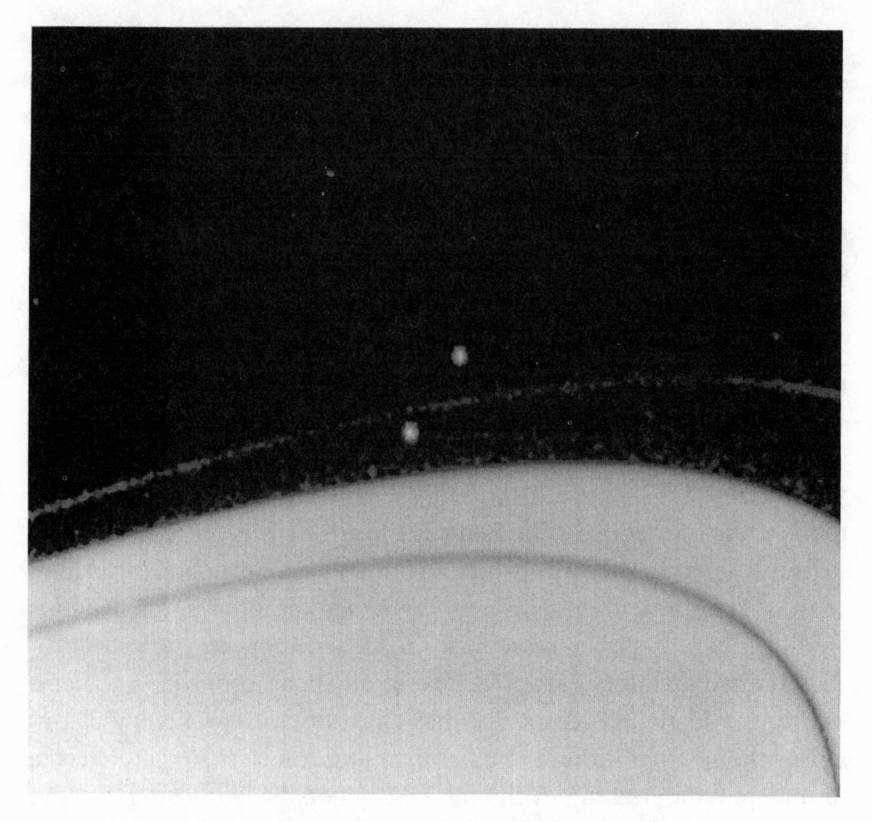

The intricate structure of Saturn's spectacular rings testifies to the remarkably complicated effects that gravity can induce. Here, two tiny satellites (Prometheus and Pandora), each measuring only about 50 kilometers across, bracket Saturn's F ring. The gravitational effects of these two shepherd satellites apparently help confine the particles in the F ring to a band about 100 kilometers wide. (NASA/Jet Propulsion Laboratory.)

Chaos is just one of a rich array of intricate behaviors that gravity can induce. Perhaps the best, and often the most puzzling, examples of these dynamical quirks occur in the miniature asteroid belts that encircle planets—in the stunningly beautiful rings of Saturn and the more diffuse strings of beads that decorate Jupiter, Uranus, and Neptune. Each ring feature has its own dynamical tale to tell.

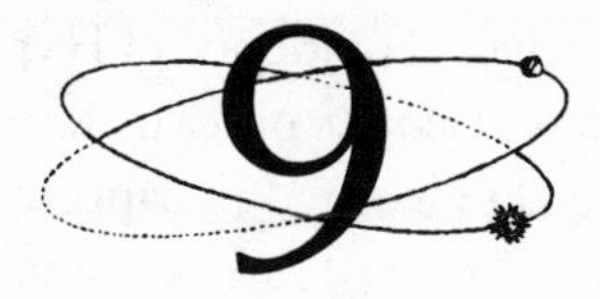

Hyperion Tumbles

•

Blazing Hyperion on his orbed fire
Still sits, still snuffs the incense teeming up
From man to the Sun's God: yet unsecure.

JOHN KEATS (1795–1821), *Hyperion*

THE BARREN Atacama Desert, a rocky strip hugging the foothills of the Andes in northern Chile, is home to tarantulas and scorpions. But the desert's bone-dry air and clear, dark nights have also attracted a different sort of visitor. Astronomers regularly trek to the clusters of telescopes perched on the hills at La Silla, Cerro Tololo, and Las Campanas. There, they take turns sampling the precious, vintage light arriving from planets, stars, and galaxies.

For three weeks in 1987, it was James Klavetter's turn. He came to the Cerro Tololo Inter-American Observatory to peer at Hyperion, a frigid, misshapen chunk of ice and rock orbiting Saturn. Three years earlier, Jack Wisdom, Stanton Peale, and François Mignard had boldly predicted that Hyperion tumbles in its orbit instead of spinning smoothly like a top. Theirs was a startling assertion because no other major natural satellite in the solar system shows such irregular behavior. Klavetter's task at Cerro Tololo was to look for erratic changes in Hyperion's brightness as a sign of tumbling.

The quest for unequivocal evidence of Hyperion's chaotic motion took Klavetter to three telescopes on two continents over a period of three and a half months. He learned to capture Hyperion's faint glint against its giant parent's glare. He overcame unforeseen technical difficulties and balky instruments, begged and borrowed telescope time to keep intact his string of nightly observations, and fretted through the bad weather that occasionally interrupted his work. In the end, he amassed enough data to present a plausible portrait of Hyperion's enigmatic demeanor and behavior.

Until the *Voyager 2* spacecraft swept past Saturn in August of 1981, astronomers had little to say about Hyperion. First identified in 1848, this small, unremarkable satellite is less than a tenth the size of Earth's moon. It orbits Saturn once every 21.28 days, staying an average of 1,480,000 kilometers away from the planet. Although its elongated orbit lies well outside of Saturn's famous rings, Hyperion feels the gravitational tug not only of Saturn but also of Saturn's huge satellite Titan, which follows an orbit just inside Hyperion's.

When *Voyager 2* began snapping pictures of Hyperion, scientists monitoring the Saturn encounter were astonished to see an extended blob instead of a sphere. At first they compared the satellite's shape to

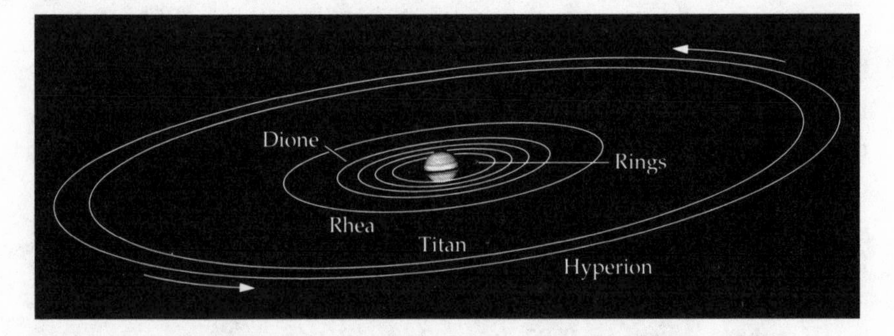

Hyperion swings around Saturn in an orbit that lies well outside Saturn's famous rings and just beyond Titan's orbit. Staying about 1.48 million kilometers from Saturn, Hyperion takes 21.3 days to circle the planet. (After data supplied by the U.S. Naval Observatory, © SERC.)

a hamburger, but subsequent views of different sides kept them revising the comparison, from a fat, oval hamburger to a deformed, battered football to a potato or a peanut. Careful analysis of the images later revealed this oddly shaped, pock-marked satellite to be nearly twice as long as it is wide, measuring approximately 380 by 290 by 230 kilometers. It is, indeed, the most irregularly shaped of the major satellites.

Stranger still, however, was its orientation. Such an elongated satellite, with its long axis pointed toward Saturn's equator, ought to be spinning about its short axis. The *Voyager* images hinted that Hyperion might instead be tilted and its spin axis skewed.

Whatever the orientation of its spin axis, Hyperion also seemed to be spinning once every 13 days while orbiting Saturn about once every 21 days. That, too, was unusual. The moon, for example, completes one rotation in the time that it takes to go once around Earth, thus always presenting the same face to an Earth-based observer. Indeed, all major satellites except Hyperion exhibit 1:1 resonances between their orbital and rotational periods.

The reason for this behavior lies not in chance but in tidal effects caused by the gravitational force exerted on a satellite by its host planet and on the planet by its satellite. Such tides arise because a planet occupies a certain amount of space. Parts of the planet located at

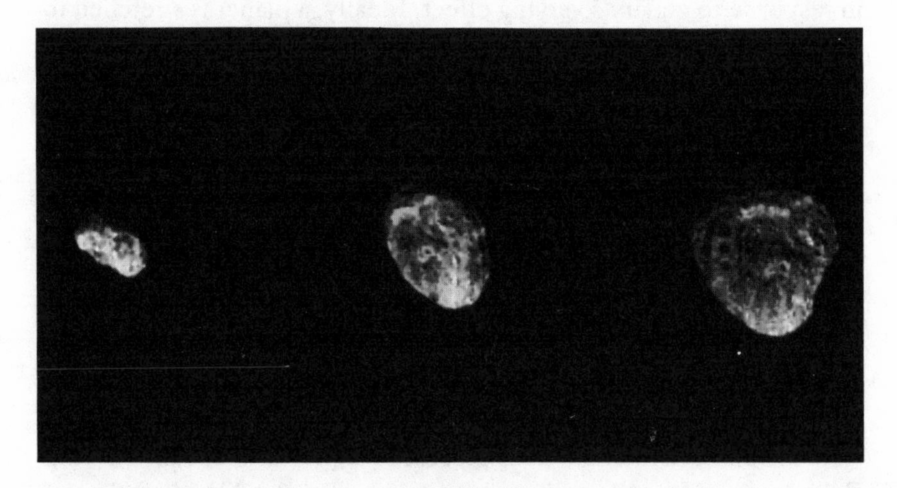

Images captured by the *Voyager 2* spacecraft provide three different views of Hyperion's irregular shape. (NASA/Jet Propulsion Laboratory.)

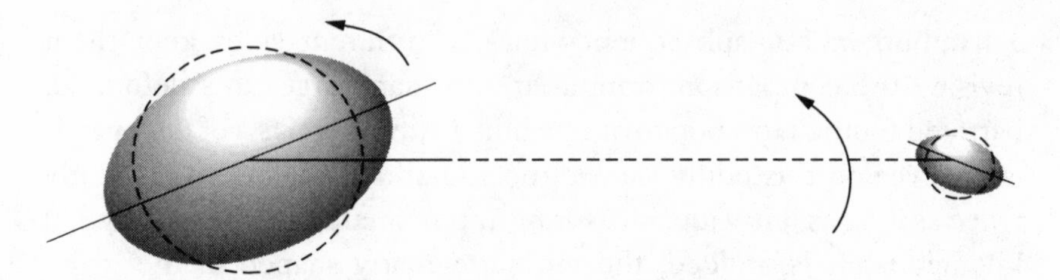

In this example, a planet's shape *(left)* is distorted by its satellite *(right)* into an ellipsoid. The long axis of the ellipsoid does not point directly at the satellite, however, because the planet's rapid rotation carries it out of alignment. Losses of energy due to tidal friction, caused by fluid motion or shifting rocks inside the planet, ensure that the planet stays out of alignment. Thus, the bulge slightly nearer the satellite feels the satellite's gravitational attraction more strongly than the other bulge. This imbalance over long periods of time gradually slows the planet's rotation. At the same time, the satellite drifts into a more distant orbit.

slightly different distances from a perturbing satellite experience somewhat different gravitational forces. But because the planet remains intact, these parts are not entirely free to move separately. Instead the planet is deformed, developing internal strains and stresses that supply the forces necessary to make all its parts accelerate together in response to gravity's varying effect. Ideally, a planet is stretched to produce a bulge that reaches its maximum extent along a line connecting the planet's center to that of the satellite, and this bulge rotates at the same rate as the satellite orbits the planet. Satellites undergo similar distortions.

Inevitably, some of the energy that goes into shifting solid or liquid material to establish the bulge is lost. Cracking rocks, breaking waves, and other irreversible processes dissipate a tiny portion of the system's energy. These effects considerably delay the planet's response to its satellite. The delay causes the planet's tidal bulge to rotate from beneath the satellite that generated it, and the gravitational attraction between this asymmetrically situated bulge and the satellite acts to slow the planet's rotation. This process even now is gradually slowing Earth's spin rate and sending the moon into a steadily more distant orbit. Another consequence of this tug-of-war is that tidal friction arising from the resultant bending and stretching damps the satellite's motion so that it, too, slows down, eventually spinning at a rate equal to the time it takes to orbit the planet.

But why hadn't Hyperion settled down, locking one face toward Saturn? Wisdom, Peale, and Mignard decided there were two key factors that prevented Hyperion from going into a synchronous state: its unusual, elongated shape and the influence of Titan, its giant satellite neighbor.

Wisdom likens the tidal process affecting Hyperion to the tossing of a partially filled, cylindrical bottle. Even though a full bottle will happily spin for a relatively long time about its long axis (which runs down the middle of the bottle from top to bottom), a partially filled bottle will not maintain such a motion. Instead, it gradually twists around so that it ends up spinning about its short axis (which runs through the middle from one side of the bottle to the other). The motion of the fluid inside the partially filled bottle introduces a kind of tidal friction analogous in effect to the tidal forces exerted on Hyperion by Saturn.

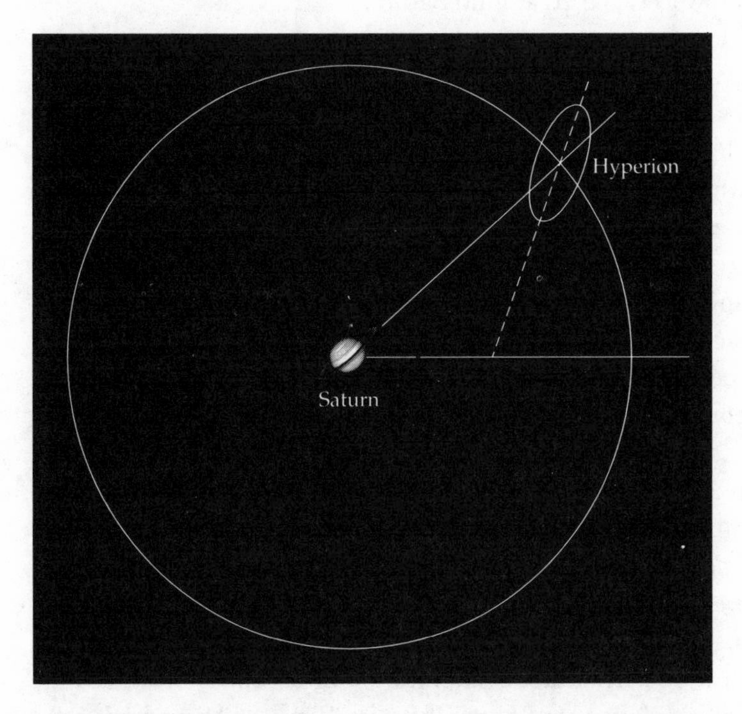

This sketch shows the orientation of an idealized Hyperion in its orbit around Saturn.

But forces acting asymmetrically don't quite explain the matter. For example, Jupiter's elongated moon Amalthea rotates stably on its shortest axis while pointing its long axis toward Jupiter.

That's where Titan comes in. Titan's considerable influence has stretched Hyperion's orbit so that it is no longer approximately circular. Because an orbiting satellite travels faster when it's closer to its host planet, Hyperion undergoes considerable changes in speed. Moreover, for every four times Titan goes around Saturn, Hyperion makes the journey only three times. As a result, the giant satellite's repetitive tugs have locked Hyperion into an apparently stable orbit.

All these influences add up to a persistently precarious balance of forces. Wisdom, Peale, and Mignard predicted that in such a situation, Hyperion's spin period should vary and the orientation and position of its spin axis should shift in only a few orbital periods. In other words, like an acrobat in slow motion, the satellite ought to be tumbling in space, even as it follows a stable orbit. Wisdom estimated that Hyperion's spin rate could go from no rotation at all to one rotation every 10 days in as few as two trips around Saturn.

To get a better handle on Hyperion's antics, Wisdom and his colleagues used a computer to solve the equations of motion describing Hyperion's spinning behavior. From the results they plotted the same kinds of phase-space diagrams that Poincaré had used in his investigations of three-body dynamics. In Hyperion's case, the diagrams showed the satellite's orientation versus its spin rate once per circuit given a variety of initial conditions. It was like taking a sequence of flash pictures timed so that Hyperion was caught at the same place in its orbit each time.

If Hyperion had a stable orientation, successive points would fall at precisely the same spot in such a diagram. If its spin axis shifted irregularly, however, then the plotted points would wander across a certain area, and Hyperion's motion could be classified as chaotic. Thus, combinations of starting points and velocities that lead to widely dispersed points, taken together, constitute a chaotic zone. In this case, instead of erratic changes in the shape or orientation of the satellite's orbit, chaos manifests itself as abrupt shifts in Hyperion's attitude.

For an irregularly shaped satellite on an eccentric orbit, chaotic zones cover a wide range of possible orbits. Thus, as Hyperion's orbit evolves, the satellite can easily arrive at the conditions necessary to

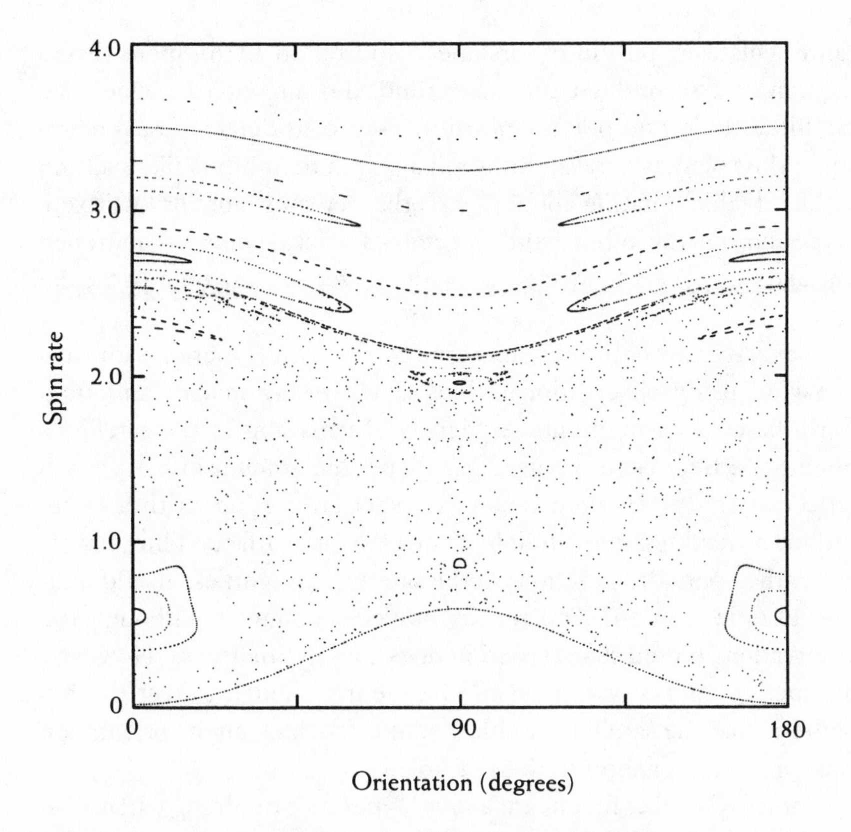

Integrating the equations of motion for Hyperion, then plotting the satellite's orientation versus spin rate each time the satellite arrives at a particular point in its orbit around Saturn reveals large regions showing a random scatter of dots, indicating chaotic behavior. The closed loops—islands of stability—show regions of regular behavior. (Courtesy of James Klavetter.)

send it into a chaotic zone. The slightest deviation of its spin axis from its normal position perpendicular to the satellite's orbital plane can then grow so rapidly that, in a few orbital periods, the axis may even end up parallel to the orbital plane before shifting to a new orientation. Locked in orbital resonance with Titan, Hyperion presently sits in just such an extended chaotic region.

The scenario developed by Wisdom, Peale, and Mignard strongly indicated that Hyperion displays a curious amalgam of the orderly and the chaotic. Because it follows a regular, predictable orbit, astronomers can calculate its position years ahead and be off by no more than a fraction of a second. At the same time, the direction in which the satellite's three axes point—its attitude—is considerably less predictable. Most planets and satellites roll along in their orbits like balls on a billiard table, rotating about an axis that remains more or less fixed. Hyperion, on the other hand, resembles a fat, somewhat flattened sausage. It appears to swing randomly, never settling on a particular axis.

Several groups of researchers tried to pin down Hyperion's unusual behavior, using observational clues from *Voyager* images and from Earth-based measurements, especially of variations in the satellite's brightness. Because of its irregular shape, the amount of light it will reflect at any given time depends on its orientation, with a larger surface reflecting more sunlight than a smaller surface. Thus, erratic rather than periodic variations in the satellite's brightness would indicate chaotic rotation. The results proved ambiguous, although the observations hinted that Hyperion does vary in brightness. However, the measurements weren't made frequently enough to test the hypothesis that the satellite tumbles, erratically changing its orientation and spin rate over short periods of time.

James Klavetter first heard about Hyperion's tumbling when Wisdom gave a talk at MIT in 1983. The notion was so intriguing that he embarked on a project to systematically monitor the satellite's brightness. He estimated that Hyperion's brightness could vary, in astronomical terms, by as much as half a magnitude. Such an effect would be easily observable with a small to moderately sized telescope.

However, Klavetter's first set of observations, made in 1984, proved inconclusive. Establishing that Hyperion's motion was truly chaotic wasn't as straightforward as Klavetter had thought, and his project turned into a full-fledged investigation. Computer models of the satellite's motion suggested that Klavetter would need a string of nightly observations extending over at least 13 weeks, a period in which Hyperion completes 4.5 circuits of Saturn. But fierce competition among astronomers generally prevents anyone from getting such a long stretch of telescope time at a major facility. Nonetheless, by offering to help with projects of interest to other astronomers, Klavetter managed to arrange a schedule that included three weeks at Cerro Tololo, a bridging week at the Lowell Observatory in Flagstaff, Arizona, and an extended run of nearly three months at the nearby McGraw-Hill Observatory.

With Saturn in the southern part of the sky, Cerro Tololo was ideally placed for observing Hyperion. Unfortunately, Klavetter found that the only photometer to which he had access for measuring Hyperion's brightness was inadequate. The problem was Saturn. Nearly half a million times brighter than Hyperion and only 1.3 to 4 arc minutes away, Saturn flooded the satellite's neighborhood with extraneous light, increasing Hyperion's apparent brightness. It was necessary to block as much of Saturn's light as possible from entering the telescope's aperture but still capture enough light to extract Hyperion's faint glint. The available equipment just wasn't good enough to meet this requirement, and in the end Klavetter had to discard the Cerro Tololo data.

At the Lowell Observatory, Klavetter had the use of a solid-state CCD (charge-coupled device) camera with which to measure both Hyperion's and the sky's brightness, but the weather didn't cooperate. He got only two noisy measurements on widely separated nights during his week-long observing run. The poor quality of the data combined with the nine-day gap before the start of observations at the McGraw-Hill Observatory led him to discard these results as well.

Luckily, Klavetter's lengthy sojourn at McGraw-Hill proved far more productive. He was able to use a fairly large, 2.4-meter telescope equipped with a good light detector and a similarly equipped 1.3-meter telescope. By scrounging time from other scheduled observers, trading nights and half-nights, and tracking comets on the side, he managed to keep his string of observations going with only a few interruptions. His persistence and ingenuity paid off.

Klavetter observed Hyperion from May 31 to August 5, obtaining 37 nights' worth of usable data over a period of 53 days. The clear weather was the best on record, with only 11 nights that were too hazy or cloudy for brightness determinations; three other sessions were ruined by equipment failures, high winds, and interfering light from Titan. Klavetter's work came to an end with the approach in August of Arizona's "monsoon season." He managed only one more observation 11 days after his first string.

Initially, Klavetter didn't know whether Hyperion's brightness would vary through the night. When he had sufficient observing time, he therefore made as many as nine independent determinations of the satellite's brightness during a single session. The brightness in fact remained virtually unchanged within this short period and the precaution proved unnecessary, but Klavetter didn't know that at the time and had to make all the measurements he could.

Klavetter recalls his long nights that summer at the McGraw-Hill Observatory as being both lonely (often he had only his dog Athabasca for company) and a test of his endurance. Both days and nights were filled with the minutiae of operating huge, occasionally balky telescopes. But the unusually long period that he spent at the observatory probably brought him closer in spirit to his astronomical forebears than

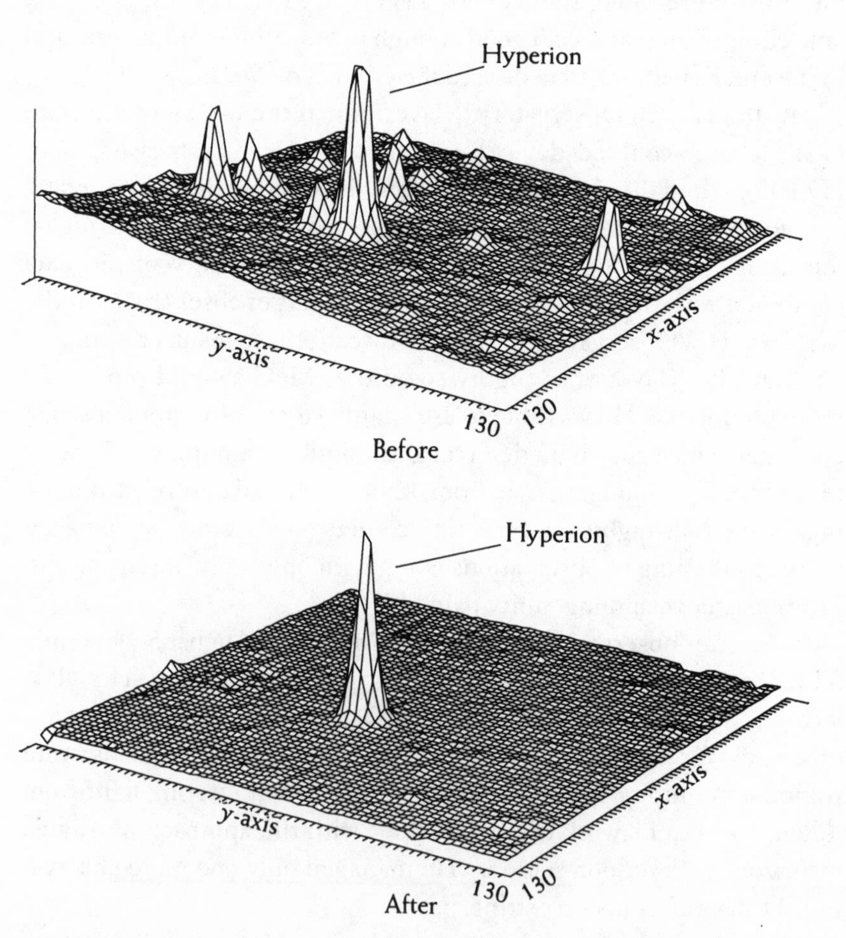

To obtain Hyperion's true brightness, Klavetter had to correct for the sky's uneven glow and subtract the images of surrounding stars. (Courtesy of James Klavetter.)

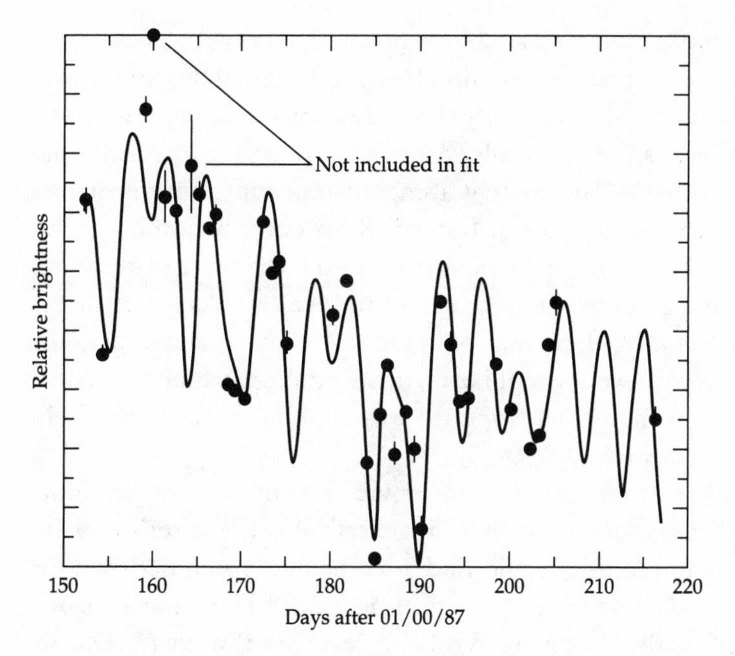

Plotting Hyperion's brightness each night for a period spanning more than 60 days produced an erratic pattern that no single periodic curve would fit. (Courtesy of James Klavetter.)

is usual in modern astronomical research, which typically means working for brief intervals with automated, remote-operated telescopes and instruments.

With a remarkably complete set of nightly observations in hand, Klavetter now began the lengthy process of analyzing the data. First he had to correct the brightness measurements for the sky's uneven glow, caused mainly by Saturn's intense glare and complicated by Hyperion's proximity in the sky to the Milky Way. Moreover, each observation of Hyperion included not only the satellite itself but also as many as a dozen stars within the same tiny fragment of sky, each one bright enough to cause more than a 1 percent difference in the final value of Hyperion's true brightness, or astronomical magnitude. To isolate how Hyperion's rotation affects its appearance, Klavetter had to take into account the satellite's position in its orbit relative to the sun and to Saturn, which also influenced its brightness at any given moment.

In the end, Klavetter was able to plot a light curve consisting of a set of points, each one representing Hyperion's actual brightness on a given night. These data certainly showed strong brightness variations, but the values varied erratically, rising and falling so irregularly that there was no obvious way to draw a smooth, repeating, wavelike curve that came near all the points. Indeed, Klavetter saw nothing in his analyses that hinted at regular behavior. No smooth curve corresponding to rotational periods from one hour to seven weeks fitted all the points satisfactorily. Hyperion could not be in any regular or periodic rotational state. Klavetter could explain the erratic variations in brightness only by assuming that Hyperion was tumbling chaotically. Nothing else he tried worked.

It wasn't enough, however, to provide essentially circumstantial evidence that Hyperion was behaving erratically. Klavetter's remaining task was to see if he could find a reasonable match between his observational data and the theoretical portrait of Hyperion's chaotic behavior originally drawn by Wisdom, Peale, and Mignard. Did its apparent motion fit the definition of a chaotic trajectory? Could theory and numerical simulation produce a light curve similar to the one observed?

Hyperion's brightness, as viewed from Earth, depends primarily on the satellite's shape. Looking more like a peanut than the slightly flattened sphere by which celestial bodies such as Earth are characterized, Hyperion naturally reflects varying amounts of light toward Earth. To properly take these factors into account, Klavetter extended the idealized scheme developed by Wisdom, Peale, and Mignard to model the satellite's dynamical behavior. He smoothed its vaguely irregular peanut shape into a flattened ellipsoid characterized by three axes of different lengths. Any slice parallel to these axes would reveal an elliptical cross section. Using largely geometrical arguments, Klavetter then hoped to establish how the direction in which this ellipsoid pointed would affect its brightness as viewed from Earth. If its axes remained fixed in certain directions, the ellipsoid's brightness would change in a characteristic repetitive fashion as it traveled along its 21-day orbit; the same pattern of rises and falls in intensity would occur again and again.

When attempting to fit a smooth curve to a set of data, researchers generally use as many observations as possible. In analyzing measurements representing an apparently chaotic system, however, Klavetter found it easier to start with a single observation and then gradually fold in additional data as the analysis proceeded. He could work backward from that one selected data point to determine the various initial conditions scattered throughout phase space that led to that particular outcome. It's a little like running a film backward to find the starting point of a ball ricocheting around a billiard table. Including a second brightness measurement in his analysis reduces the number of potential trajectories. In other words, of the rotation states initially bunched together that happen to fit the first point selected from the light curve, only a few would also fit the second selected point.

Using a computer, Klavetter exhaustively searched the appropriate regions of phase space to establish that a certain set of initial conditions in his theoretical model would lead to shifts in Hyperion's axis of rotation that closely approximate the erratic motion needed to satisfy the dynamical model of a chaotically tumbling satellite. Proceeding largely by trial and error, he demonstrated step by step that the scattered data points in his light curve corresponded to just such an allowed chaotic trajectory. His repeated computer simulations of an ellipsoid's movements starting at just the right points supplied the necessary evidence.

But was the fit good enough? Chaos itself precludes a definitive, convincing answer to that question. No matter how well earlier points may match a curve, once it starts to wander from its projected trajectory in phase space, the curve's future course may shift so much that it never returns to its original path. This is precisely what makes it impossible to predict the future evolution of a chaotic system from real measurements. Any uncertainty in these measurements grows so rapidly that the result is complete ignorance. Even if it had been possible to measure Hyperion's orientation and spin direction to 10 figures at the time of *Voyager 1*'s encounter with Saturn, it would still have been impossible to predict its attitude when *Voyager 2* arrived nine months later.

The irony of the situation is that the equations of motion are known and can be solved. But the solutions show such an exquisitely fine sensitivity to initial conditions that predictability is lost.

Although Klavetter couldn't absolutely and positively prove that Hyperion's tumbling, as seen in its brightness variations, represents chaotic motion, his results provided useful dynamical information about the satellite. *Voyager 2* had photographed only about 50 percent of Hyperion at high resolution, and planetary scientists had found it impossible to draw a complete map of its cratered surface. Still, it was clear that Hyperion has a remarkably irregular shape that only roughly approximates the perfect ellipsoid that Klavetter used in his computer models. Nonetheless, Klavetter matched the satellite's light curve to better than 5 percent. This agreement between theory and observation enabled Klavetter to derive information about Hyperion itself from his brightness measurements. These data strongly hint that Hyperion has a nearly uniform density, with little likelihood of significant concentrations of mass within its body that could unbalance or skew the satellite's motion.

Interestingly, *Voyager*'s images of Hyperion's battered surface supply a modicum of circumstantial evidence supporting the notion that it resides in a chaotic zone. Hyperion may very well have started out as a more or less spherical satellite. Like other satellites of Saturn, it was probably heavily bombarded by primordial orbital debris. However, while the other satellites may have fragmented and reassembled several times over many millions of years, Hyperion's chaotic motion may have prevented that from happening. Instead of falling back onto the parent body, fragments ejected from crash landings on Hyperion remained adrift. What we now call Hyperion may be merely the biggest chunk left over from a massive, catastrophic impact that left the satellite a scarred, craggy shadow of its former self.

At the same time, it's highly unlikely that Hyperion started out in a chaotic state. In the distant past, Hyperion's rotational period, or day, was much faster than its orbital period, or year. As tidal forces over eons gradually slowed the satellite's rotation, Hyperion would inevitably have reached rotational velocities that sent it into a chaotic zone. At this critical stage, its spin axis was probably nearly perpendicular to its orbital plane. Once the satellite entered the chaotic zone, however, this hard-won attitude stability was undone in a matter of days, and Hyperion started to tumble.

Hyperion now seems trapped in this type of motion. Because the chaotic zone surrounding the 4:3 resonance between the orbital periods of Hyperion and Titan is so vast, the chance that the satellite's orbit will ever reach one of the few small islands of stability in the vicinity appears minimal. And even if Hyperion eventually manages to reach a synchronous state, it will continue to tumble because the island associated with this particular orbit itself exhibits unstable attitudes.

Of course, Hyperion isn't the only oddly shaped object orbiting a planet. Hyperion may be unique in showing chaotic motion now, but Wisdom has conjectured that all irregularly shaped satellites may have tumbled in the past at the point where they approached synchronous rotation with the host planet. For example, although the chunks of rock known as Phobos and Deimos now orbit Mars in perfectly regular orbits, long ago they may have tumbled chaotically for periods ranging from 10 million to 100 million years. Neptune's satellite Nereid, which has a highly eccentric orbit and a size comparable to Hyperion, may have gone through a similar stage.

Does any contemporary evidence suggest that a particular satellite may have evolved through a time of chaos? Some researchers have suggested that the lava flows, volcanic cones, and other exotic surface features of Miranda—a satellite of Uranus—may have resulted from a period of chaotic tumbling. But detailed analyses indicate that this is unlikely. While enhanced tidal effects during such a period could provide enough heat to generate some volcanic activity, Miranda's period of attitude instability would have occurred only as it entered its synchronous rotation state. And that would have been relatively soon after its formation, because it would take only 300,000 years or so to slow down sufficiently to reach a synchronous state. The episode of chaotic tumbling would have occurred too soon after Miranda's formation to account for the wide range of ages assigned to the satellite's surface features.

Already enshrined in textbooks, the Hyperion example stands as the first conclusive demonstration of chaotic motion in the solar system. It also demonstrates a much more general phenomenon—a surprising, hitherto unexpected, chaotic episode in the adolescence of any misshapen satellite slowed by tidal forces before it locks one face toward its host planet. Equally surprising is the way tumbling in this situation is the distinguishing mark of chaos. At various times in the history of the solar system, a number of irregularly shaped satellites undoubtedly evolved into and then out of a period of chaotic tumbling.

Only Hyperion, however, happened to drift into a situation that causes it to display chaotic tumbling to this day.

Chaos is only one of many types of dynamical behavior possible in satellites and planets. Gravity works its wonders through an extensive palette of movements not readily apparent in the differential equations that represent its effects. As Newton and his successors realized, the trouble with determining the stability and long-term history of orbits is that the differential equations involved are notably opaque. It takes a fair amount of tedious algebra to extract even a glimmering of how a

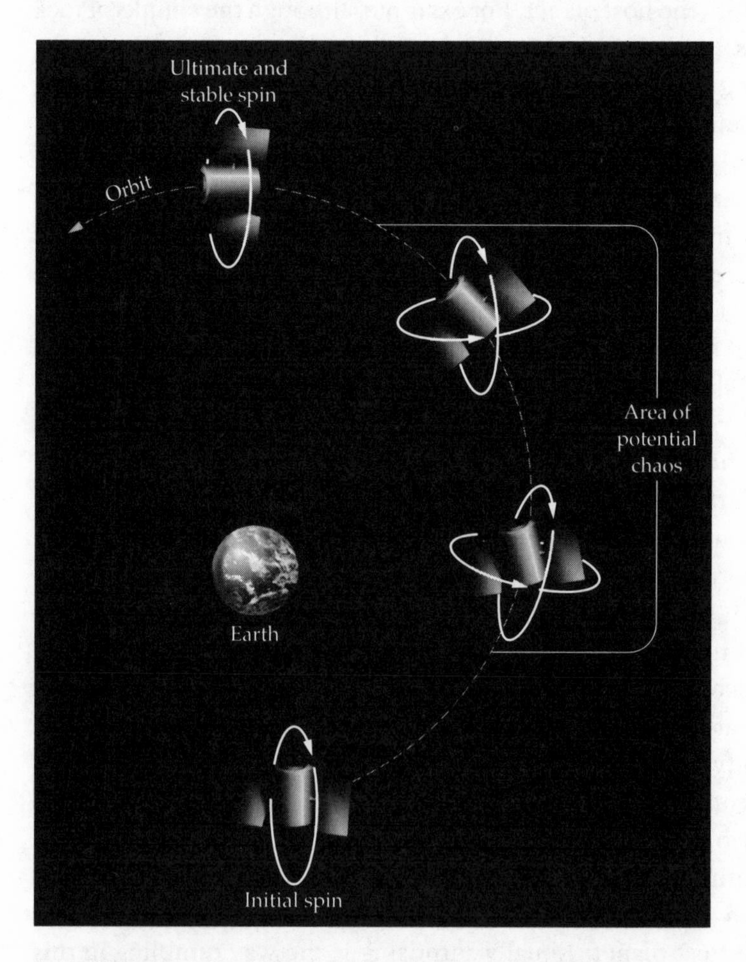

An artificial satellite spinning on its long axis gradually makes the transition to spinning on its short axis. As it moves from one axis to the other, the satellite may have to pass through a dangerously chaotic zone. (University of Wisconsin-Madison.)

planet or a moon, let alone an artificial satellite in Earth's uneven gravitational field, behaves. In particular, identifying within a differential equation those features that are responsible for long-term behavior requires a means of cleanly separating crucial, long-term trends from a background of short-term fluctuations.

The work of Charles-Eugène Delaunay in the middle of the nineteenth century represents one especially heroic attempt to characterize the moon's orbit by separating long- and short-term trends. Educated as an engineer, Delaunay taught engineering and had a professional interest in both astronomy and mechanics. At that time, no engineer's education was complete without a course in celestial mechanics, and Delaunay even wrote a treatise on astronomy specifically for engineers. But his private passion, which he pursued methodically for two decades starting in 1847, concerned the slight inconsistencies that had marred Laplace's attempt to account fully for the moon's movements.

Before the computer age, scientists like Delaunay could envision no way to predict the motion of a celestial body other than by expressing the approximate solutions of the relevant differential equations as lengthy strings of algebraic expressions in so-called power series. Once they had found a suitable set of formulas, they would turn to the burdensome task of breaking the method down into a sequence of convenient steps for calculating specific answers. Developing the formulas and then restructuring them into lengthy tables that clerks in almanac offices could use to compute lunar or planetary positions required years of effort. And if a clerk of modest skill couldn't perform the calculations necessary to predict the moon's change in position in less time than it took the moon to make the journey, that effort, no matter how elegant the method and accurate its outcome, had to be considered a failure.

Working alone, Delaunay dedicated 20 years to his pursuit of the moon. Rigorous to a fault, he took great pains to document his hand calculations, even showing how each individual term arose from one long chain of algebraic terms to the next. After 10 years spent carrying out his program and another 10 spent checking the work, Delaunay finally published the results of his stupendous algebraic manipulations and calculations in two volumes, in 1860 and 1867. But even these interminable tables, which allowed the moon's position at a given time to be calculated with greater precision than ever before, failed to match the accuracy of observational data from ancient Greece.

A hundred years elapsed before anyone had the interest or capability to check Delaunay's work. In the 1950s, mathematical astronomers started to investigate Delaunay's methods for possible application to the computation of the orbits of artificial satellites. At the same time, researchers were developing special computer programs that worked directly with the abstract symbols of algebra rather than merely with numbers. These novel computer-algebra systems provided astronomers with just the tools they needed to check Delaunay's lists of formulas and algebraic expressions, which filled page after page of his two hefty volumes.

André Deprit, Jacques Henrard, and Arnold Rom were among the first to take this approach in revisiting Delaunay's long-neglected work. In 1970, at the Boeing Scientific Research Laboratories in Seattle, the trio used a computer-algebra system they had developed to run through Delaunay's formulas and tables. Their calculation, which essentially duplicated Delaunay's effort, required about 20 hours of computer time, a minuscule amount compared with the 20 years that Delaunay had lavished on the project. Remarkably, they found only three errors in the entire work, all in rather insignificant terms. Moreover, two of the errors were consequences of the first, in which Delaunay slipped up by using the fraction $^{23}/_{16}$ instead of $^{33}/_{16}$ as the coefficient in front of a particular algebraic term.

Emboldened by this success, Deprit and his colleagues extended Delaunay's work to include additional terms, thereby achieving a significantly higher level of precision than previously possible in computing both lunar positions and the orbits of artificial satellites. By modifying the theory slightly, they could also take into account the effects of atmospheric drag and Earth's nonspherical shape on satellites in low-altitude orbits.

But such methods are generally too cumbersome, tedious, and complicated for use by engineers who want to get a quick overview of which Earth orbits may be suitable for a particular space mission. Variations in Earth's gravitational field caused by the planet's bulging equator, the uneven distribution of mass within its rotund body, and other distortions strongly influence an artificial satellite's orbit. To keep it from tumbling unnecessarily or drifting away from its planned la fonction R ne contient plus aucun terme périodique; elle se trouve donc réduite à son terme non périodique seul, terme qui, en tenant compte des parties fournies par les opérations 129, 260, 349 et 415, a pour valeur

$$R = \frac{\mu}{2a}$$

$$+ m'\frac{a^2}{a'^3}\Big\{\frac{1}{4} - \frac{3}{2}\gamma^2 + \frac{3}{8}e^2 + \frac{3}{8}e'^2 + \frac{3}{2}\gamma^4 - \frac{9}{4}\gamma^2 e^2 - \frac{9}{4}\gamma^2 e'^2 + \frac{9}{16}e^2 e'^2 + \frac{15}{32}e'^4 - \frac{33}{2}\gamma^4 e^2$$

$$+ \frac{9}{4}\gamma^4 e'^2 + \frac{75}{16}\gamma^2 e^4 - \frac{27}{8}\gamma^2 e^2 e'^2 - \frac{45}{16}\gamma^2 e'^4 + \frac{45}{64}e^2 e'^4$$

$$+ \Big(\frac{9}{16}\gamma^2 + \frac{225}{64}e^2 - \frac{27}{16}\gamma^4 - \frac{387}{32}\gamma^2 e^2 + \boxed{\frac{23}{16}\gamma^2 e'^2} - \frac{225}{128}e^4 + \frac{825}{64}e^2 e'^2 + \frac{9}{8}\gamma^6$$

$$+ \frac{3897}{64}\gamma^4 e^2 - \frac{99}{16}\gamma^4 e'^2 - \frac{1431}{256}\gamma^2 e^4 - \frac{1419}{32}\gamma^2 e^2 e'^2 - \frac{225}{512}e^6 - \frac{825}{128}e^4 e'^2\Big)\frac{n'}{n}$$

$$- \Big(\frac{31}{32} - \frac{33}{8}\gamma^2 - \frac{971}{32}e^2 + \frac{465}{64}e'^2 + \frac{273}{64}\gamma^4 + \frac{5709}{64}\gamma^2 e^2 - \frac{117}{4}\gamma^2 e'^2 + \frac{4989}{256}e^4$$

$$- \frac{1905}{8}e^2 e'^2 + \frac{3255}{128}e'^4\Big)\frac{n'^2}{n^2}$$

$$- \Big(\frac{255}{32} - \frac{31515}{1024}\gamma^2 - \frac{551115}{4096}e^2 + \frac{6885}{64}e'^2 + \frac{20511}{512}\gamma^4 + \frac{927831}{2048}\gamma^2 e^2$$

$$- \frac{218115}{512}\gamma^2 e'^2 + \frac{1622985}{16384}e^4 - \frac{4069635}{2048}e^2 e'^2\Big)\frac{n'^3}{n^3}$$

$$- \Big(\frac{5515}{192} - \frac{296779}{3072}\gamma^2 - \frac{6380965}{12288}e^2 + \frac{16285}{24}e'^2\Big)\frac{n'^4}{n^4}$$

$$- \Big(\frac{28841}{288} - \frac{113818307}{294912}\gamma^2 - \frac{1681901051}{1179648}e^2 + \frac{1393609}{384}e'^2\Big)\frac{n'^5}{n^5}$$

$$- \frac{9814775}{36864}\frac{n'^6}{n^6} - \frac{428268199}{663552}\frac{n'^7}{n^7}$$

$$+ \Big[\frac{9}{64} - \frac{45}{16}\gamma^2 + \frac{45}{64}e^2 + \frac{15}{128}e'^2$$

$$+ \Big(\frac{225}{512} - \frac{1935}{256}\gamma^2 + \frac{7425}{1024}e^2 + \frac{225}{64}e'^2\Big)\frac{n'}{n} + \frac{869}{512}\frac{n'^2}{n^2} - \frac{10391}{8192}\frac{n'^3}{n^3}\Big]\frac{a^2}{a'^2}\Big\}.$$

This page from the second volume of Charles Delaunay's massive hand calculation of the moon's position as a function of time to a precision never before attained shows the only major error that a modern recomputation revealed in his work. The fraction in the boxed expression should be 33/16 instead of 23/16. (Reprinted with permission from Richard Pavelle, Michael Rothstein, and John Fitch, "Computer Algebra." *Scientific American*, December 1981, p. 151.)

course, aerospace engineers must constantly monitor and adjust a satellite's position and orientation. Otherwise, over time, gravitational imbalances can shift an orbit or rattle a satellite sufficiently to prevent it from fulfilling its mission. In fact, a satellite's working lifetime in orbit depends in large measure on the amount of fuel it expends on maneuvers to correct orbital deviations imposed by irregularities in Earth's gravitational field.

Indeed, putting a satellite into a stable orbit around Earth is no simple matter. At the dawn of the space age in 1957, no one had any detailed knowledge of precisely where, or even whether, a satellite might drift over the course of its lifetime in space. Aerospace engineers concentrated on the intricacies of tracking these primitive sputniks, and they learned through experience that orbits could change quite dramatically. It was not uncommon to lose track of a satellite or two for periods ranging from hours to months.

Consider a weather satellite placed in an equatorial orbit so that it always hovers over the same spot on Earth. As it travels from west to east, the satellite matches its motion with that of the planet thousands of kilometers below. But slight variations in Earth's gravitational field tend to pull the satellite away from its designated orbit, and engineers must send signals that trigger the firing of bursts of gas to shift its altitude.

Generally, an equatorial orbit is sufficiently stable that it takes minimal fuel to keep a satellite in place for long periods. When left to itself, however, a satellite launched into an equatorial orbit will slowly but inexorably drift to a position over the Indian Ocean, where Earth's gravity is slightly stronger than elsewhere because of the presence of an unusually large concentration of dense material somewhere beneath the ocean. Over the years, many satellites and much orbital debris have ended up in this orbital graveyard.

Increasingly, engineers and mission planners want to find stable orbits to reduce the need for on-board fuel and constant monitoring and correction. Furthermore, certain orbits have turned out to be more useful than others, so precise knowledge of the long-term behavior of orbits is necessary to prevent collisions on these popular skyways. It is essential for aerospace engineers to know precisely where such stable orbits lie and how much leeway they have.

To help aerospace engineers select appropriate orbits, André Deprit, who had joined the National Bureau of Standards (now the

National Institute of Standards and Technology), collaborated in the late 1980s with a trio of scientists at the Naval Research Laboratory—Shannon Coffey, Liam Healy, and André's son Étienne—to develop a flexible and convenient technique for visualizing how distortions of Earth's gravitational field influence the dynamics of Earth-orbiting satellites. Coupling graphics with a powerful computer capable of performing many calculations simultaneously, they were able to draw a special kind of contour map of phase space—the abstract space whose dimensions are position and momentum—that linked a satellite's movements with Earth's gravitational field.

To obtain a global picture of long-term trends in a satellite's motion, the researchers used an advanced version of the same kind of computer-algebra tools developed earlier by Deprit and his colleagues when they tackled Delaunay's lunar formulas. These techniques allowed them to isolate long-term trends, which are generally of more immediate concern to planners of space missions than the small, repetitive shifts and wobbles that jostle a satellite minute by minute and day by day in its orbit.

The product of this work was no ordinary map of Earth's surface. Instead it depicts a two-dimensional landscape whose coordinates represent not longitude and latitude but an orbit's angular momentum, or rotational motion, and the angle between the equator and the line joining Earth's center to the point at which the satellite is nearest Earth (the "argument of perigee"). Thus, each point in the resulting "phase portrait" represents a single orbit with a certain angular momentum and orientation; a particular color or shade of gray is assigned to each point according to the energy of the specific orbit.

In the resulting maps, strips of different colors portray how Earth's uneven gravitational field affects orbits. If Earth were perfectly spherical and no other perturbations disturbed its gravitational field, the portrait would be all one color and every orbit would be stable. In such a situation, an orbiting satellite traces out an ellipse, with Earth's center at one focus. Such orbits range in eccentricity from circles to highly elongated ellipses, and the curves may lie in the same plane as Earth's equator or at any other inclination—including one at right angles to Earth's equatorial plane, which would put a satellite into orbit over both poles.

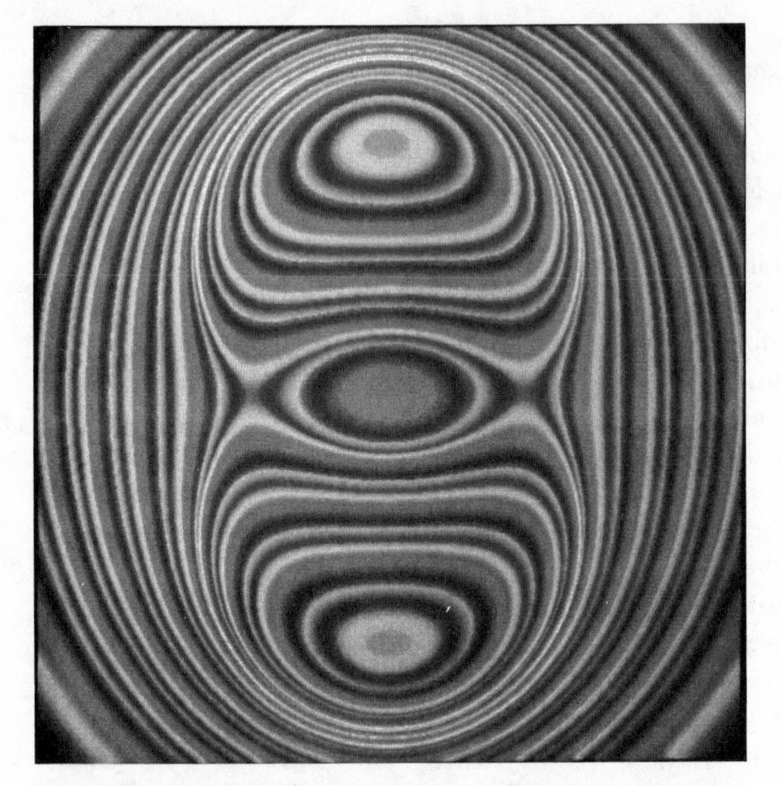

In this view of the "northern hemisphere" of the phase space corresponding to the equations of motion for satellites in Earth orbit, the central point represents a circular satellite orbit. Points farther from the center signify orbits with larger eccentricities and smaller inclinations. (Liam Healy, Étienne Deprit.)

Maps corresponding to such gravitational effects as those induced by Earth's somewhat asymmetric equatorial bulge comprise separate sets of concentric rings woven together in intricate, startling geometries at the common borders, creating a geography of steep peaks, shallow valleys, and narrow passes. Certain features on these maps point to orbits that tend to remain stable or have minimal wobble; others indicate strange orbits that induce flips, oscillations, or aimless drifting.

With the help of these computed portraits, engineers can trace the evolution of all possible trajectories for a wide range of orbital shapes and positions. By examining a map's contours, they can readily identify families of orbits that tend to remain stable. Having located the coordinates of an orbit of interest, they can solve the underlying differential equation to obtain a more precise idea of how a satellite in that orbit would behave. Finally, using a little elementary mathematics, they can work out the precise conditions needed to lift a given craft into a temporary parking orbit and the velocity needed to send it into its final orbit.

At first, Coffey and his colleagues applied their visualization technique to a simplified set of equations that included only deviations caused by Earth's equatorial bulge. These equations applied best to a satellite in a low orbit, where the moon's influence would not be great and variations in Earth's mass distribution would average out. Using such a stripped-down model, the researchers obtained important insights into the nature of orbits near the so-called critical inclination of 63 degrees.

The equations of motion at that particular angle are notoriously difficult to solve, and before the late 1950s no one knew what an orbit at that inclination would be like. So it came as a shock when the Soviet Union launched a number of satellites into orbit at the critical inclination, dramatically demonstrating that satellites in such orbits tend to stay put. It didn't take long for the United States to follow the Soviet lead, and in recent years as many as two-thirds of the U.S. Navy's satellites have gone into orbit at this inclination, mainly to simplify the chore of controlling and tracking them.

The Navy was particularly interested in launching clusters of satellites into such orbits—say, five satellites at equal angles from one another—on an ellipse. When Navy engineers attempted to maintain such an array, however, they had trouble tracking some of its members. The satellites drifted more than they were supposed to.

Using phase portraits to visualize orbits near the critical inclination, Coffey and his co-workers uncovered the true complexity of the types of motion possible under these dynamical conditions. Some orbits are stable, and some clearly are not. Later, the researchers added the effects of much smaller perturbations to get results useful to engineers planning satellite orbits between 500 and 3000 kilometers above Earth's surface. An additional step allowed them to take into account the perturbations caused by the moon and sun.

In a paper published in 1990 in the journal *Science,* Coffey and his colleagues noted how their maps revealed new dynamical phenomena, which in turn prompted detailed scrutiny of the equations used to perform the computations in the first place. They wrote: "We have now come full circle. Analytical study of a dynamical system prompted graphical representations to support our results. Improvements in the visualization techniques revealed new phenomena, which brought us to refine our mathematical analysis."

In the miniature solar systems that ring many of the planets, it's apparent that satellites, both artificial and natural, can display multifarious instabilities, some of which can be attributed to chaos. But what about the planets themselves? Are their motions truly predictable and forever determined, or are they too subject to the whim of chaos? As the next chapter shows, arriving at even a tentative answer to this question requires computations on a scale that would have amazed even Johannes Kepler.

Digital Orrery

•

Chance, too, which seems to rush along
with slack reins,
is bridled and governed by law.

BOETHIUS (ca. 480–525),
The Consolation of Philosophy

THE LABYRINTHINE complex of laboratories and offices at the Massachusetts Institute of Technology houses an enormous number of time-and-space machines. Fueled by electrons and piloted by mathematical logic, these marvelously versatile electronic vehicles permit far-ranging, eon-straddling excursions into the universe. Of these machines, one is charged with a particularly audacious, daunting task.

Launched in the spring of 1991, this new vessel bears a mundane, technically correct name: the Supercomputer Toolkit. It is the fruit of a unique collaboration between Jack Wisdom, who earned recognition for his pioneering explorations of the dynamical reasons for gaps in the asteroid belt and for the erratic gyrations of Saturn's satellite Hyperion, and Gerald Jay Sussman, an affable and animated computer scientist with interests ranging from artificial intelligence to astronomy. Both have played an influential role in bringing computation to the forefront of questions of stability and chaos in celestial mechanics.

Like the Argonauts of ancient myth, these two MIT professors have set out on a daring, perilous course into uncharted waters. Their custom-built supercalculating prodigy, barely the size of a modest filing cabinet, may hold the key to the solar system in its silently shuttling electrons. One of its main functions is to solve the equations governing the motions of the solar system's nine planets in order to predict their future and uncover their past.

Of course, Wisdom and Sussman are not the first to tackle the question of the solar system's long-term future and its stability. These unresolved, excruciatingly subtle issues have plagued celestial mechanics for at least 200 years. Indeed, the first major attempt to settle the matter takes us back to France during the latter part of the eighteenth century and the astonishingly resilient career of Pierre-Simon Marquis de Laplace.

To the self-assured youth born 20 years earlier on a Normandy farm, Paris in 1769 glowed like a grand stage on which to strut his mathematical prowess. Armed with polite recommendations from a number of influential people and testimonials from professors at the University of Caen (where he had studied for five years), Laplace was confident that he would soon gain entry to the intellectual circles and fashionable salons of the burgeoning city. He arrived at the door of Jean d'Alembert, the aging scientist, mathematician, and philosopher who after a long career, which included the studies of the moon's motion described in Chapter 6, had become the dominant figure in French academic life. A Parisian famed for his brilliant conversation, d'Alembert was also known for his willingness to sponsor and aid young aspirants.

When Laplace arrived seeking an interview, a servant carried in his letters of recommendation. D'Alembert refused to see the visitor, sending back a brusque message that he was not interested in young men who came only on the recommendation of prominent people. Momentarily dismayed but detecting an opening, the young mathematician returned to his lodgings and proceeded to compose a lengthy letter concerning the general principles of mechanics. The ingenious reasoning and evident mastery of the subject matter reflected in the letter captured d'Alembert's attention, and he invited the young man to call. In his reply, d'Alembert wrote, "Sir, you see that I paid little enough attention to your recommendations; you don't need any. You have introduced yourself better. That is enough for me; my support is your due."

Within a few days, the youth received an appointment as a professor of mathematics at the École Militaire. By the time he died in 1827, Laplace had more than justified d'Alembert's confidence in his considerable talents. He had also proved a wily, calculating opportunist who had managed to survive the bloody transition from the old regime to republican France, the birth and demise of Napoleon's empire, and the subsequent restoration of the monarchy. Deftly shifting with the political winds, Laplace continually escaped imprisonment and exe-

Pierre-Simon de Laplace (1749–1827). (Smithsonian Institution.)

cution, despite his open association with such eventual victims as the chemist Antoine Lavoisier. Whether as a member of a commission charged with reforming the nation's system of weights and measures or as a founding member of the Bureau des Longitudes, he was always at hand to serve and assist whoever happened to be in charge.

Laplace had a lifelong knack for being in the right place at the right time. He even served for six weeks in 1799 as minister of the interior under Napoleon. But his record as an administrator was so dismal that Napoleon commented near the end of his life, in exile on the desolate island of St. Helena, that Laplace "saw no question from its true point of view; he sought subtleties everywhere, had only doubtful ideas, and finally carried the spirit of the infinitely small into the management of affairs." Aware of the value of mathematics and science to statecraft, however, Napoleon continued to hold Laplace in high esteem and eventually honored him with a title. In the bitter propaganda wars between Britain and France that occupied much of the Napoleonic era, it didn't hurt that Laplace, with his extensive studies in celestial mechanics, could with some legitimacy style himself as France's Newton.

Existing alongside these political adventures was an undeniably brilliant gift for mathematics and its application to astronomy. Starting in 1773, Laplace devoted most of his scientific life to explicating the role of Newtonian gravitation in the grand scheme of the solar system. With considerable skill and virtuosity, he deployed mathematics as a precise tool with which to tackle, and in some cases solve, a number of significant puzzles in solar system dynamics.

The issue of the solar system's stability was a particularly formidable problem. The tangle of mutual gravitational interactions exhibited by the known planets and the sun was so complex that no complete mathematical solution seemed possible. Newton himself had noted certain irregularities in the movements of the planets that he suspected could lead to the disruption of the solar system unless orbits were, in effect, reset at strategic moments. He concluded that divine intervention was periodically necessary to maintain the system's equanimity.

In the seventeenth and eighteenth centuries, the perplexing movements of Saturn and Jupiter furnished a particularly noteworthy example of such apparently equilibrium-destroying tendencies. Observations compiled over many decades indicated that Jupiter's orbit was continuously shrinking, whereas Saturn's was slowly but steadily expanding. By 1625 Johannes Kepler had already sensed that there was something wrong, and Edmond Halley had explicitly introduced corrections for these movements in the tables of planetary positions he prepared in 1695. His adjustments amounted to a deviation over a thousand years of nearly 1 degree in Jupiter's computed position and more than 2 degrees in Saturn's.

A short time after his arrival in Paris, the young Laplace began a concerted effort to improve the calculation and characterization of planetary orbits. As a first step, he proved mathematically that irregularities in the eccentricities and inclinations of planetary orbits are bounded. According to his theoretical analysis, the values of these orbital parameters oscillate about fixed values rather than growing larger and larger. In other words, although the planets follow paths that regularly deviate from ideal orbits, they don't stray very far.

The application of Laplace's mathematical model of an idealized solar system, along with new observations, helped to solve the Jupiter-Saturn riddle. First, a careful reexamination of seventeenth- and eighteenth-century observations revealed that the changes in these orbits were the opposite of what had been assumed. In reality, Jupiter's orbit was expanding, whereas Saturn's was shrinking. Then, in 1784, Laplace demonstrated that because Saturn makes two circuits of the sun in roughly the time that Jupiter makes five, the planets meet at nearly the same place along the ecliptic every 59 years or so. Over time, these coincidences accumulate into perceptible gravitational effects that displace the planets sufficiently to alter their observed positions in the sky. Because the periods are not expressed precisely as a ratio of small whole numbers, the position of conjunction slowly shifts, and the apparent reciprocal effect of the two planets reverses after about 450 years. Thus, the observed perturbations in these two orbits are simply manifestations of a slow oscillation with a period of 900 years.

On his success in using Newton's law of gravitation to settle this matter, Laplace would later note: "The irregularities of the two planets appeared formerly to be inexplicable by the law of universal gravitation—they now form one of its most striking proofs. Such has been the fate of [Newton's] brilliant discovery, that each difficulty that has arisen has become for it a new subject of triumph, a circumstance that is the surest characteristic of the true system of nature."

By 1786, and again using his idealized mathematical solar system, Laplace had proved that any deviations in the eccentricities and inclinations of planetary orbits are small, constant, and self-correcting. He concluded that these perturbations could not accumulate to wreak havoc on the solar system's arrangement. Because inclinations and eccentricities remain within well-defined limits, the solar system's configuration would remain essentially unchanged forever. Like a huge, self-adjusting clockwork driven by the universal force of gravitation, the solar system is inherently stable and predictable—according to Laplace's theory.

Attracted to any mathematical idea that could help interpret nature, Laplace brought together and codified a wide range of techniques and results that shed light on the workings of the solar system. Published between 1799 and 1825, the five thick volumes of his monumental *Traité de mécanique céleste* (Treatise on celestial mechanics) not only introduced the term *celestial mechanics* but also offered a splendid demonstration of what could be achieved by treating all motions in the solar system as purely mathematical problems. This massive work embodied Laplace's great desire to transform science into a pursuit fueled and guided by mathematics alone.

Displaying considerable confidence in his approach and methods, Laplace proudly introduced his definitive treatise with this ringing declaration: "Towards the end of the seventeenth century, Newton published his discovery of universal gravitation. Mathematicians have, since that epoch, succeeded in reducing to this great law of nature all the known phenomena of the system of the world, and have thus given to the theories of the heavenly bodies, and to astronomical tables, an unexpected degree of precision. My object is to present a connected view of these theories, which are now scattered in a great number of works. The whole of the results of gravitation, upon the equilibrium and motions of the fluid and solid bodies, which compose the solar system, and the similar systems, existing in the immensity of space, constitute the object of Celestial Mechanics. . . . Astronomy, considered in the most general manner, is a great problem of mechanics. . . . The solution of this problem depends, at the same time, upon the accuracy of the observations, and upon the perfection of the analysis."

To Laplace, all of nature functioned like the solar system as a clockwork. Given the initial position and motion of every particle in a system, he argued, the future course of that system—whether a collection of planets or a falling apple—was entirely determined and, in principle, calculable. In his classic statement on determinism, Laplace stated: "Assume an intelligence that at a given moment knows all the forces that animate nature as well as the momentary positions of all things of which the universe consists, and further that it is sufficiently powerful to perform a calculation based on these data. It would then include in the same formulation the motions of the largest bodies in the universe and those of the smallest atoms. To it, nothing would be uncertain. Both future and past would be present before its eyes."

Laplace's prestige, celebrity, and popular writings brought this deterministic view before a large audience. But in the wide-ranging discussions and passionate debates that followed, philosophers, mathematicians, and other scholars raised a variety of objections. For example, small causes could initiate large effects. Imagine a rock precariously poised on the tip of a mountain peak. Toppled by the slightest shove, the rock could easily trigger a massive avalanche in the course of its descent down the mountain's slope.

Did analogous instabilities exist within the solar system? Laplace didn't think so, but other mathematical astronomers who followed in his footsteps weren't so sure. They were starting to worry about resonances and their potentially unpredictable influence on planetary orbits. Even Laplace, when he tried to tame the moon's motion, failed to account fully for all the details of its orbit. He was left with a nagging discrepancy between theory and observation that wouldn't go away.

As noted in Chapter 7, it took Henri Poincaré's keen insight and genius to notice the uncertainties built into the mathematics that describe even so simple a system as three bodies moving under the influence of their mutual gravitational attraction. However, what this astonishingly cantankerous mathematical behavior had to say about the stability of the real solar system proved to be a question of considerable delicacy. The multifarious gravitational interactions among the planets continually modify planetary orbits. Ellipses stretch, shrink, and wobble. But so long as the changes occur in more or less regular cycles, these quasi-periodic orbits remain stable. Although planets in such orbits don't precisely repeat their movements at regular intervals—as would a planet in a solar system consisting only of itself and a central star—their wanderings can be expressed as the net result or sum of an ensemble of such periodic movements.

However, if an orbit's eccentricity, size, or inclination begins to vary erratically, this irregular behavior serves as a barometer of possible instability. In the most dramatic case, such variations could cause one planet to cross the orbit of another, leading to a catastrophic crash or to the hurling of one planet out of its regular orbit, and perhaps out of the solar system. Although astronomers have traditionally assumed that a planet's excursions from regular motion have been small in the past and will remain so in the future, the fact that dynamical systems can unexpectedly display chaos alongside regular behavior suggests that the same phenomenon could turn up in the solar system.

In 1963, in an extension of Poincaré's work, Vladimir I. Arnol'd provided a crucial mathematical insight into the question of stability. As mentioned in Chapter 7, he proved that any solar system, despite its potential for chaos, will for all practical purposes remain quasi-periodic, and hence stable, provided that the masses, inclinations, and eccentricities of the planets are sufficiently small. But does our particular solar system fit his criteria for stability? Is it on an essentially quasi-periodic trajectory, or are its past and future truly obscured by the uncertainty implicit in chaos?

Of course, the answer can be learned simply by letting the physical experiment run its course. But with 4.5 billion years already expended and at least that many more years to go, the experiment runs far too slowly for the matter to be settled one way or the other. Mathematics provides the shortcut, but only if the equations really do capture the solar system's crucial elements and only if they can be solved with the necessary precision.

In principle, a sufficiently powerful computer can step through the necessary calculations, but numerical integrations of the solar system take extraordinarily large amounts of computer time. Any direct calculations have to proceed in increments small enough to follow each planet. Speedy Mercury revolves around the sun in 88 days, whereas distant Pluto requires centuries. Moreover, because nothing much happens in mere thousands of years, the evolution of orbits must be tracked for many millions of years to yield meaningful insights into their long-term behavior. No single computational model readily accommodates such a tremendous range of time scales without wasting a great deal of effort on calculations that individually don't amount to much.

It took mathematical astronomers many years to develop the computational instruments they needed to peer into the solar system's future. At first, even the most advanced computers proved inadequate to the task, and researchers had to compromise by using simplified solar systems, studying special cases, or limiting the time spans covered by their numerical integrations. The 1970s saw several separate investigations into the motions of Jupiter, Saturn, Uranus, Neptune, and Pluto. The inner planets were notoriously difficult to include because such short intervals were required to compute the evolution of their orbits accurately, and hence they were nearly always neglected. Still, integration of even a truncated solar system couldn't reach very far into the past or future.

Those investigators lucky enough to take advantage of the scant and minutely scheduled computational resources that were gradually becoming available for such large-scale projects could make limited forays into the solar system's future or past and watch orbits evolve over eons. However, because very precise integrations over simulated periods spanning millions of years remained prohibitively expensive, many researchers stayed on the sidelines. By 1983, no one had traveled more than five million years into the solar system's future, and those who had managed to get this far saw no clear signs of irregularity.

But there was an alternative approach, which Gerald Sussman came to personify. Sussman arrived at MIT in 1964—a rather insecure, 17-year-old freshman—where he encountered the freshly minted, challenging world of computation, full of opportunities for exploration and discovery. Throughout his undergraduate days, Sussman applied his electronics and computer programming expertise to various projects, including several that involved robotics, at the university's nascent artificial intelligence laboratory. In 1973, as part of his doctoral dissertation in mathematics, he spent a large chunk of time developing the provocative notion that errors, or "bugs," have what he called a "virtuous" character.

Describing the basis for his theory of "problem solving by debugging almost-right plans," Sussman asked, "How much time has each of us spent tracking down some bug in a computer program, an electronic device, or a mathematical proof? At such times it may seem that a bug is at best a nuisance, or at worst a disaster. Has it ever occurred to you that bugs are manifestations of powerful strategies of creative thinking? That, perhaps, creating and removing bugs are necessary steps in the normal process of solving a problem? Recent research at the MIT AI Laboratory indicates that this is precisely the case." Out of this theory came Sussman's intricate and acclaimed computational model of skill acquisition. It was embodied in a computer program called HACKER, which could, for example, figure out what moves are required for a robotic arm to retrieve a particular block from a jumbled heap. This problem-solving system had the ability to improve its performance with practice.

The years following Sussman's first experiences with mainframe computers—those hulking giants of the digital world—saw an explosion in the number and types of computers available to researchers. Personal computers and workstations became commonplace. No laboratory or office was complete without one or more of these machines to collect data, massage information, or pass on messages. At the opposite extreme, a succession of supercomputers were claimed to be the world's fastest number crunchers, and researchers with huge problems to solve sought access to these comparatively scarce, expensive machines.

But these glamorous supercalculators had the deficiencies of any device designed to be all things to all people. To achieve high levels of performance while maintaining flexibility, designers were forced to make difficult choices, and the resulting compromises often impeded the solution of particular problems. No general-purpose computer was ideally suited to a given application.

Sussman came to believe that many researchers, instead of focusing on the physics or chemistry of the problem at hand, were expending too much effort on the computational subterfuges needed to convert software elephants into sleek thoroughbreds. He argued that the time and energy spent tweaking computer programs so that results could be efficiently extracted from a general-purpose computer would be better spent collaborating with computer engineers on made-to-order machines tailored for specific computational problems. Custom-built machines could exploit a problem's idiosyncrasies and special patterns in ways not available to an off-the-shelf computer.

Sussman developed a strategy of building special-purpose hardware when there was a scientific advantage in making such an effort.

Early on, he had designed and built two special-purpose microprocessors for handling a dialect of the computer programming language LISP. His interest in hardware found another focus in Jack Wisdom's fascinating discoveries about asteroid orbits (described in Chapter 8). With his long-standing appreciation of astrophysics, Sussman was skeptical of the shortcuts Wisdom had taken in his calculations and of the approximations on which he had built his model of asteroid movements. Sussman wanted a more definitive answer, and he proposed the design and construction of a special-purpose computer whose sole aim would be to calculate the behavior of a small number of bodies moving in roughly circular orbits under Newtonian gravity. Not only would such a custom-built machine provide the answers Wisdom needed, it would do it for far less money than it cost to program and run a supercomputer.

Sussman called his proposed machine the Digital Orrery, named for the marvelously complex mechanical and clockwork models of the solar system constructed during the eighteenth and nineteenth centuries. These intricate simulacra, consisting of spheres of various sizes

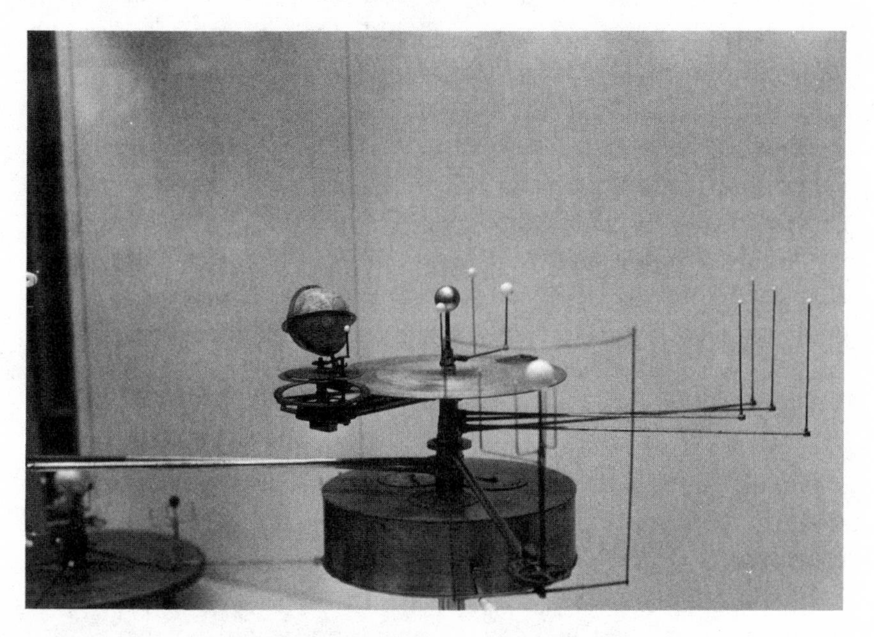

Orreries—mechanical models of the solar system—were concrete manifestations of the popular seventeenth-century belief that the universe behaves like a well-regulated clock.

mounted on gear-driven axles with extended arms, quite rigidly and somewhat vaguely mimicked the movements of planets and satellites.

Sussman's machine was to be a low-cost, speedy computer specifically configured for solving problems in celestial mechanics. But it wasn't easy to obtain funding to build a special-purpose computer. Sussman recalls, "In spite of the obvious advantages of a special-purpose machine, that notion that one could be designed and built was simply not part of the cultural outlook of the astrophysics community. In fact, a preliminary proposal for constructing the Digital Orrery to the astronomical instruments division of the National Science Foundation was rejected on the grounds that the project would be infeasible. Such an attitude is striking in view of the immensely complex engineering projects that have been accomplished by this same community when building telescopes. Building computers is just not that difficult."

Finally, in 1984, Sussman saw an opportunity to test his ideas; on sabbatical at Caltech, he brought together a team to design and build his planetary time-and-space machine. In the space of a year, six people—three theoretical physicists, two computer scientists (including Sussman), and a technician—assembled about a cubic foot's worth of electronic components into a unique computer the size of a hefty shoebox.

In fact, the Digital Orrery was actually 10 independent computers, each one wired up on a separate circuit board. Instead of having one central processor plodding through all the necessary computations one at a time, this machine coordinated the work of 10 processors, dividing up tasks among them to achieve greater efficiency. Wired together like an electronic solar system to simplify and speed up operations, the computer itself mimicked the structure typical of a problem in celestial mechanics, with a processor assigned to each planet and the sun.

The machine also contained special circuitry to accelerate and streamline the few basic arithmetical operations that the Digital Orrery would have to perform over and over again to solve problems in celestial mechanics. The silicon chips had been designed by Hewlett-Packard engineers for use in the company's computers, and Sussman's friends there helped him out. "The Orrery would not have been possible without them," Sussman later said. Using these chips to wire in certain mathematical operations permanently, Sussman was able to

The Digital Orrery, sitting on the grass at Caltech in 1984, was a special, custom-built computer for carrying out high-speed, high-precision investigations of the dynamics of the solar system. (Courtesy of Gerald Jay Sussman.)

reduce the overhead that normally results from the way a conventional computer must fetch from memory the instructions necessary to perform these operations every time a program calls for one.

These special features and refinements added up to a diminutive machine that could calculate motions in the solar system at about one-third the speed of a Cray-1 supercomputer and 60 times the speed of a VAX 11/780, a popular scientific research computer about the size of a closet. Consuming and dissipating little more than the power of a 150-watt light bulb, the Digital Orrery could race through the solar system's past and future at an astonishing rate. Unlike a mechanical orrery geared to unchanging, perfectly predictable orbits, this new electronic mechanism had the flexibility to calculate, one step at a time, what a planet's motion would be—no matter what the outcome.

Ironically, by the time Sussman arrived at Caltech to build his machine, Wisdom had already left. But when Sussman returned to MIT the following year with his new, custom-built celestial calculator in hand, he discovered that Wisdom had just been hired as an assistant professor of planetary science. It was inevitable that the two would begin to collaborate. As a first test, Wisdom borrowed the machine to check his hypothesis that chaotic asteroid orbits could lead to certain gaps in the distribution of asteroids between Mars and Jupiter. Both the Digital Orrery and Wisdom's hypothesis passed with flying colors. The fledgling machine was ready to tackle a much more formidable problem.

The Digital Orrery's first major excursion carried it about 100 million years into the future and 100 million years into the past. The power of mathematics combined with specialized digital circuitry allowed the machine to complete this remarkable journey in only a few months, thereby compressing a significant chunk of the solar system's history into a human time scale. Taking into account the movements around the sun of Jupiter, Saturn, Uranus, Neptune, and Pluto, the simulation provided the first true insight into the long-term future of the outer solar system. Stretching across 200 million years, it represented a hundredfold improvement in the period covered by previous integrations.

Of course, there was no way of actually picturing where each planet was at any given moment. Instead, Wisdom, Sussman, and their col-

In this photo taken in 1988, Jack Wisdom *(left)* and Gerald Sussman *(right)* sit beside the Digital Orrery. (Donna Coveney/MIT.)

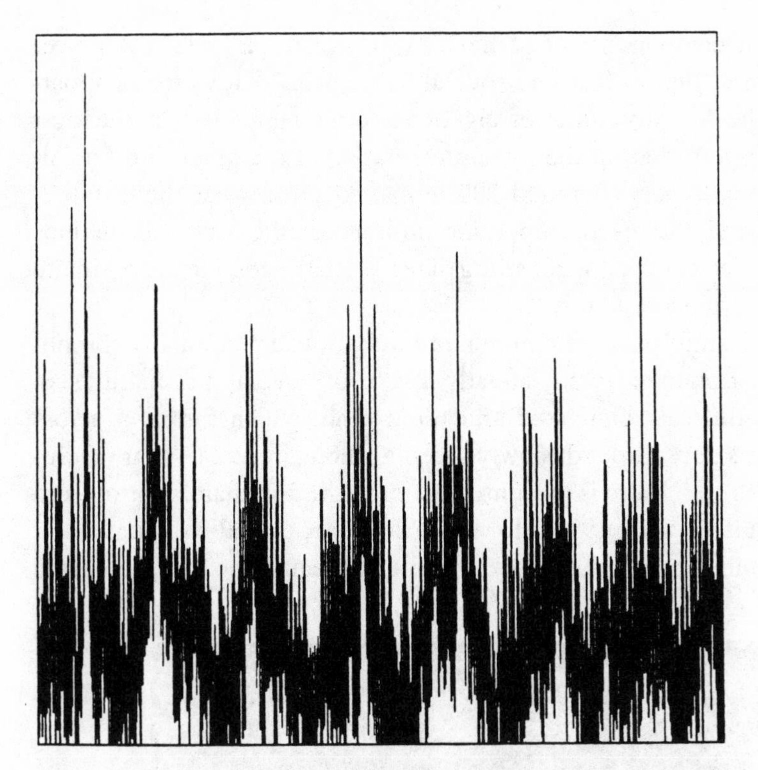

The power spectrum of a variable related to the eccentricity of Jupiter's orbit reveals the complexity of the planet's motion. A lone planet following a purely periodic orbit would have a single characteristic frequency, shown by a single vertical line in its corresponding power spectrum. (Courtesy of Jack Wisdom.)

laborators concentrated on tracking statistically changes in key orbital parameters such as eccentricity, inclination, and semimajor axis. From these data, they could produce a variety of so-called power spectra, which highlight the particular frequencies of the repetitive patterns characterizing a planet's orbit. In such plots, a lone, perfectly formed planet orbiting the sun would show up as a single spike, indicating that it travels in a precisely periodic orbit at a characteristic frequency corresponding to the time it takes to orbit the sun. More complicated interactions exhibit several characteristic frequencies, which appear as a succession of spikes in the corresponding power spectrum.

The power spectra churned out by the Digital Orrery spotlighted the inadequacy of conventional perturbation methods for capturing important components of planetary orbits. For example, the power spectrum of Jupiter featured several frequencies that were not apparent in the lengthy sums of algebraic terms contained in the best available perturbation theory approximation of the planet's orbit. Although this theory included 200 terms, covering what theorists believed were the most important influences, the new calculations showed that an even longer string of terms was needed to approximate Jupiter's orbit adequately.

In the simulation, the motion of Pluto proved particularly complicated. Astronomers were already aware of several peculiarities in Pluto's unusually elongated and tilted orbit, which carries it across Neptune's path. Indeed, slowly creeping along in its 248-year revolution of the sun, Pluto is presently nearer to the sun than Neptune and will remain so until 1999. However, these occasional crossings don't lead to collisions or even close encounters, because the two planets are

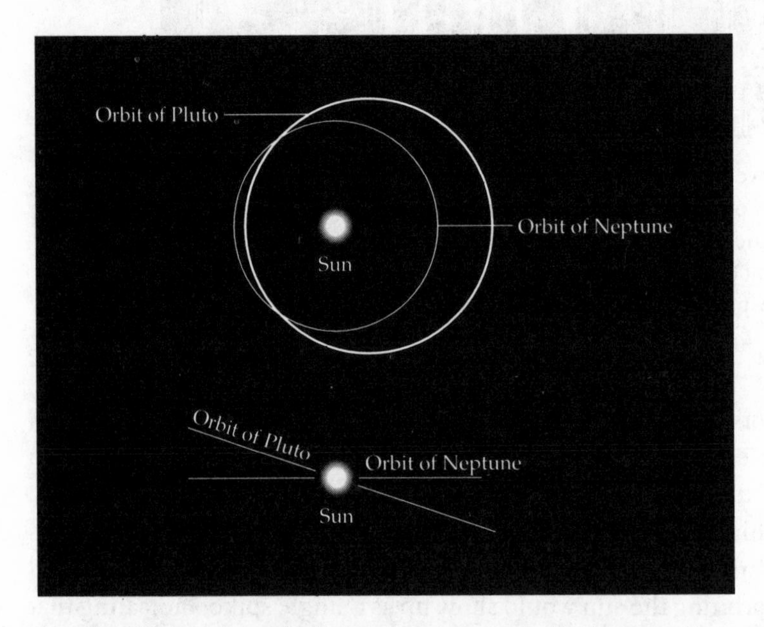

Pluto's orbit is more elliptical and more steeply inclined to the plane of the ecliptic than is the orbit of any other planet in our solar system. In fact, its orbit is so eccentric that Pluto is occasionally closer to the sun than Neptune is. Furthermore, Neptune completes three laps around the sun in the time it takes Pluto to do two, but Neptune always passes Pluto at a point where the two orbits are farthest apart.

in an orbital resonance. Neptune makes roughly three circuits of the sun in the time it takes Pluto to make two. Thus, whenever Neptune overtakes Pluto, the latter is at the far end of its ellipse—as far away from the sun and from Neptune's path as possible—and Neptune is always distant from Pluto when a crossing occurs.

Moreover, earlier numerical integrations had revealed an additional, more subtle resonance. Pluto's perihelion (the longitude at which it is closest to the sun) and its ascending node (the longitude at which its orbital plane crosses that of Earth) are locked together. The gradual shift in the locations of these events has exactly the same period.

The Digital Orrery's calculations unveiled several new features of Pluto's motion. The planet shows a surprisingly large number of strong variations that repeat themselves over millions of years, something that no algebraic perturbation theory or numerical integration carried to fewer than a million years would readily reveal. The most suggestive sign of possibly erratic variations in Pluto's orbit was the general noisiness of its power spectrum, indicating the presence of a plethora of resonances and other interactions that persistently jiggle the planet.

To identify signs of chaos, the researchers performed separate simulations aimed just at Pluto, looking at what happens when two Plutos start at slightly different points. But these calculations failed to establish that the two trajectories would diverge significantly and go their own ways, as a chaotic system requires.

Wisdom and Sussman then decided that they had to perform even longer integrations to further illuminate Pluto's peculiarities. But now they had to confront head-on the small but nagging numerical discrepancies that had plagued their initial integrations. Because of its strictly limited capacity for expressing numbers to many decimal places, a computer can't easily handle the huge numbers that arise in celestial mechanics. This means that answers must be rounded off, and the resulting numerical errors, though minuscule, can easily snowball into serious deviations the further a computer carries an integration.

Thus, computations carried into the future and then reversed to return the system to its starting point generally miss their target. On these round trips in time, planets fail to return exactly to their initial positions. Moreover, round-off errors slightly change the system's total energy at every step; in fact, because the solar system neither gains nor

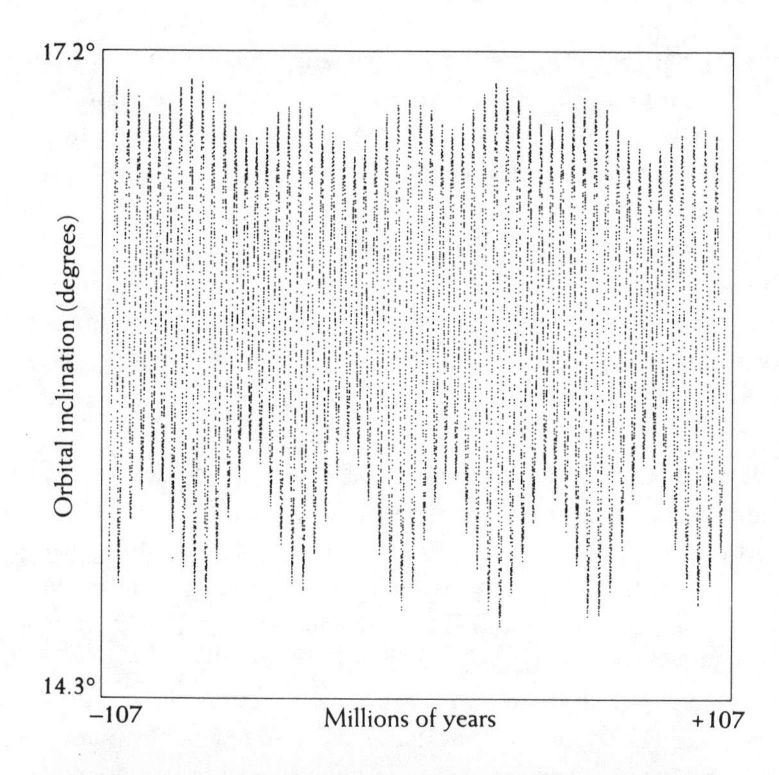

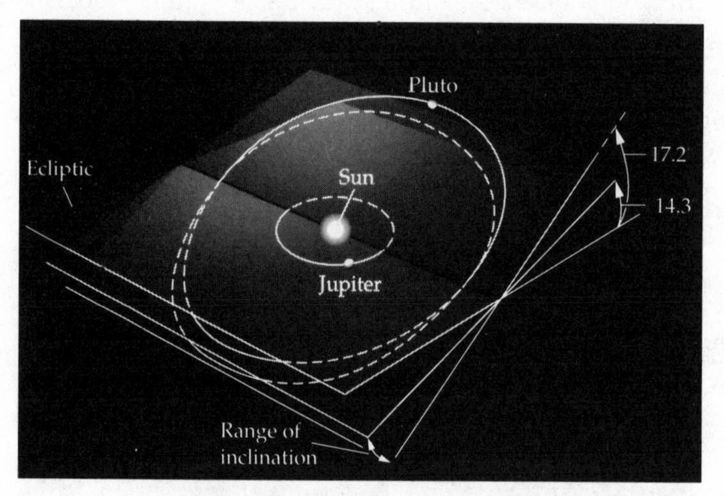

Computations spanning 214 million years show changes in the inclination of Pluto's orbit *(bottom)* that reveal a 34-million-year cycle on top of a 3.8-million-year oscillation *(top)*. (Reprinted with permission from Piet Hut and Gerald Jay Sussman, "Advanced Computing for Science." *Scientific American*, October 1987, p. 152.)

loses energy as it evolves, the system's energy should remain constant. Especially where chaos resides, with its sensitive dependence on initial conditions, round-off errors introduce uncertainties into predictions of planetary behavior in the distant future. An additional error arises from the need to approximate the continuous movements of the planets by discrete steps, equivalent to sudden jumps in position from one moment to the next.

Both types of error contributed in approximately equal amounts to the discrepancies in the Orrery's integrations, and the engineering solution was to balance the errors against each other as much as possible to minimize the net effect. The accumulation of round-off errors had limited Wisdom and Sussman's initial integrations to 100 million years, so corralling the steady growth in the energy error was crucial to permitting longer, more reliable integrations. Wisdom and Sussman had several clues with which to work. They didn't understand precisely why the energy error grew as it did in their particular machine. This was a consequence of the way the logical circuitry had been designed and the complex ways in which it interacted with the specific numbers of a calculation. But they discovered that for some mysterious reason, the rate at which the energy error accumulated appeared to depend on the size of the step from one calculation to the next in the chain of integrations.

Initially, Wisdom and Sussman had used a somewhat arbitrary step size of 40 days; that is, the orrery calculated the positions of the planets 40 days after the starting point, then used the newly computed values as the next starting point from which to calculate another set of positions 40 days further into the future, and so on. When the researchers played around with the step size, they found that the error accumulation rate increased for some step sizes but decreased for others.

At first that insight led nowhere. Then, at a conference on control theory at the NASA Ames Research Center in California, Sussman found himself seated next to William Kahan, an accomplished computer scientist and specialist in the idiosyncrasies of computer arithmetic. Sussman had never met Kahan before and was unaware of his expertise, but during a particularly boring talk he ended up describing his problem. Kahan observed that it was possible that a portion of the error was somehow being canceled out. He warned that this fluke was

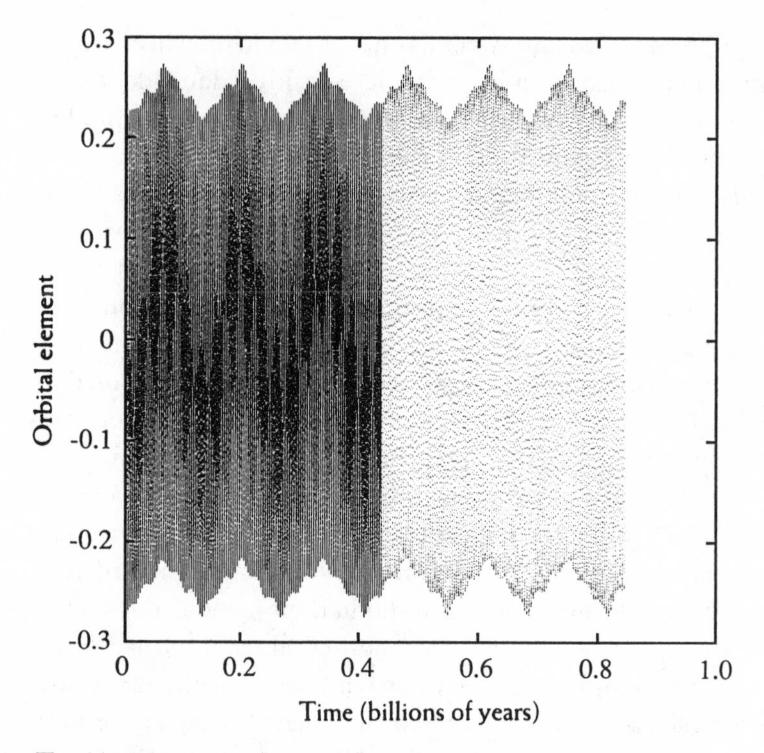

Tracking the value of a quantity related to the eccentricity of Pluto's orbit over 845 million years shows oscillations having a period of 137 million years. (Courtesy of Jack Wisdom.)

potentially dangerous and strongly urged Sussman to check it out carefully. When he got back to MIT, Sussman put his student Panayotis Skordos to work on the problem of characterizing how error depended on step size. It took Skordos about six months of testing to determine that they could safely proceed.

Astonishingly, by trial and error, they also found a step size that essentially kept the energy error from growing, no matter how long the integration. Something about the circuitry in the Digital Orrery mandated the use of 32.7-day steps, and no one could figure out why this particular number worked best. But they could demonstrate in test runs that round-trip errors in the position of Jupiter were smallest when they used this particular step size. Curiously, they also found that when this number was used, the round-trip discrepancies for all the planets were comparable, which was not true for other step sizes.

With this change, the Digital Orrery could simulate the motions of the outer planets for nearly a billion years, and for some unknown reason related to the quirks of the machine's arithmetical engine, the resulting integration would prove significantly more accurate than any previous integrations of the solar system. For example, Sussman and Wisdom could follow Pluto's orbital motion forward as many as 3.4 million revolutions about the sun and then take it backward in time to only about 10 arc minutes—one-third the apparent width of the moon—from its starting point.

Thus began the Digital Orrery's five-month, nonstop odyssey into the future of the five outer planets. Eventually reaching 845 million years, the machine managed to span roughly 20 percent of the solar system's age.

When the integration came to an end and the researchers took a look at plots of the behavior of planetary orbital parameters, they saw little that looked chaotic. The four larger outer planets all displayed the subtle variations in orbital eccentricity, inclination, and orientation already seen in early integrations. But the orbits stubbornly remained regular. Enmeshed in a number of different resonances that induced several long-term cyclic variations, Pluto's orbit again proved more complex than the rest.

Examining their accumulated data, Wisdom and Sussman could discern several new cycles in addition to those they had found in the earlier investigations (which they were reassured to see again). Among these periodic variations were 3.8-million-year and 34-million-year oscillations in the inclination of Pluto's orbit and tantalizing traces of even longer cycles spanning 150 million and 600 million years. The orbit's perihelion appeared to be caught up in cycles lasting 3.7 million, 27 million, and 137 million years.

But Pluto displayed no obvious bouncing around during the entire 845-million-year duration of the simulation. For the most part, whatever variations occurred, the planet's movements appeared regular and could just as easily be associated with stability as with instability. The evidence of chaos surfaced when Wisdom and Sussman checked the behavior of twin Plutos placed at slightly different starting points. The positions and velocities of the initially neighboring planets, pictured as trajectories in phase space, diverged at an exponential rate. Calculations of the so-called Lyapunov exponent (named for Russian mathematician Aleksandr M. Lyapunov) used to characterize chaotic behavior showed that the distance between the two Plutos would increase by a factor of *e*—roughly 2.72—every 20 million years.

This particular finding suggests that although astronomers can predict Pluto's position next year with only a small error, they cannot specify with any degree of certainty precisely where Pluto will be 100 million years from now. The planet's starting position and velocity cannot be ascertained and expressed exactly, and this small uncertainty inevitably leads to long-term unpredictability—the hallmark of chaos.

Although this sensitivity makes prediction difficult, it doesn't necessarily show that Pluto will swing into a new orbit or drift away from the solar system any time soon. It could happen, but there's no way of knowing and, of course, Pluto has apparently clung to its present course without significant deviation for a considerable portion of the solar system's history. It's possible that Pluto's present trajectory in phase space has just the right characteristics to put the planet in a relatively small chaotic zone and that the major elements characterizing its orbit remain confined to the small though still essentially unpredictable variations typical of a relatively mild case of chaos. However, if this zone happens to be connected by some narrow path to a much larger one, and if Pluto eventually meets the criteria to take this path, the planet could head off on a radically different orbit without warning, just as certain asteroids can abruptly swing out of orbits confined between Mars and Jupiter.

What this says about Pluto's past isn't entirely clear either. The fact that the planet's orbit presently appears to be chaotic supports the hypothesis that Pluto probably formed near its present orbit or at a low inclination and eccentricity and then evolved to its current oddball orbit by passing through one or more chaotic regimes. As a lone survivor from the days when the solar system was awash in small bodies, Pluto may just have been lucky enough to escape being flung out of the solar system or sent crashing into the larger planets already in existence.

However, like Laplace's mathematically idealized solar system, the Digital Orrery's model represents only an approximation of the true solar system. Wisdom and Sussman made a number of assumptions and took several shortcuts, which they hoped and believed would not materially affect their basic conclusions. Their model did not take into account such exotic effects as the occasional close approach to the solar system of a wandering star and the slow, steady dimunition of the sun's mass as it emits light and particles in a furious solar wind. They neglected the small effects of the inner planets by assuming that these bodies move so fast and are so near the sun that their gravitational effects blend in with the sun's. They also set aside the minor corrections required to take into account the refined gravitational theory embodied in general relativity.

In addition, there were unavoidable errors in the assumed masses of the planets, which are still being refined using new measurements from spacecraft and Earth-based radar systems, and in the actual positions and speeds of the planets used as starting points for the simulation. Wisdom and Sussman also assumed Pluto to be a zero-mass "test particle." In other words, their simulated Pluto responded to the gravitational pull of the other planets but itself exerted no influence.

Chaos has a way of spreading. Each planet, even one as small as Pluto, to some degree affects the others through gravitational interactions. Indeed, an international consortium of solar system dynamicists participating in a project called LONGSTOP (Long-term Gravita-

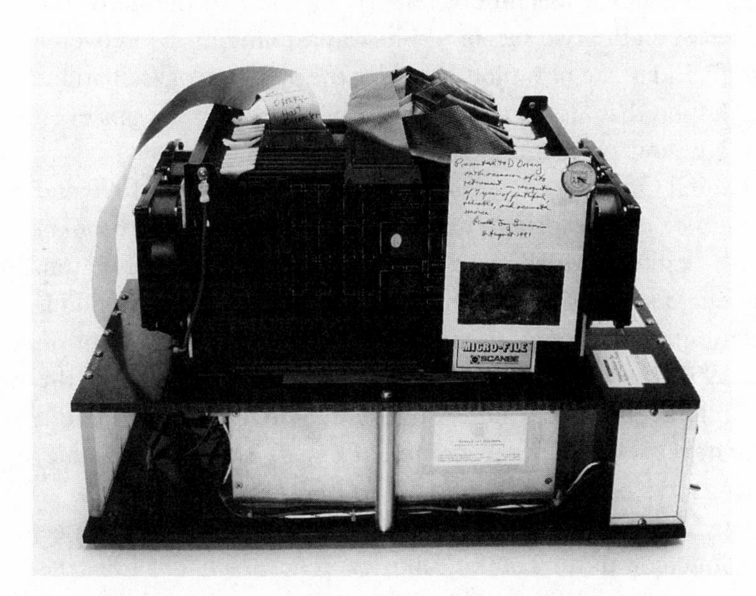

The Digital Orrery, wearing its retirement watch, now rests at the Smithsonian Institution's National Museum of American History. (Smithsonian Institution.)

tional Study of the Outer Planets) furnished some hints of curious resonances and exchanges of energy among the massive outer planets. Using a Cray supercomputer at the University of London, this group integrated the orbits of the outer planets for 100 million years. From the results, they identified a small number of variations in the orbital properties of Jupiter, Saturn, Uranus, and Neptune, but no signs of gross instabilities. Nonetheless, the orbits appear to change in a somewhat open-ended way, leaving a crack through which chaos can creep into the system.

In 1991, Jack Wisdom and his colleague Matthew Holman revisited Pluto, this time applying a modified version of the computational shortcut that Wisdom had used to track the paths of asteroids (described in Chapter 8). These computations followed the evolution of the outer planets for 1.1 billion years, making them the longest probe yet into the evolution of orbits in the outer solar system. They also provided independent confirmation that Pluto's motion is chaotic.

The new results reproduced all the principal findings of the Digital Orrery's adventure, again pinpointing cycles in Pluto's movements with periods of 34 million and 137 million years. Given the different kinds of mathematical machinery underlying the two methods, the similarity between the two sets of results was astounding. Moreover, a slight rising trend in the new plots hinted at the presence of oscillations in Pluto's motion that last more than a billion years—or perhaps even at an achingly slow, permanent shift in Pluto's orbit.

The Digital Orrery completed its final computations in 1990 and retired to a quiet, contemplative life in the archives of the Smithsonian Institution's National Museum of American History in Washington, D.C., where it joined antique mechanical orreries and other astronomical and computational relics of earlier times. Upon its retirement on August 2, 1991, Sussman affixed an official MIT gold watch to the machine and the following note: "Presented to D. Orrery on the occasion of its retirement, in recognition of 7 years of faithful, reliable, and accurate service."

Sussman and Wisdom continue their odyssey with their new machine, the Supercomputer Toolkit. Like its predecessor, it handles the mathematics of celestial mechanics with consummate ease. Unlike the Digital Orrery, however, it has the speed and capacity to encompass all the planets of the solar system.

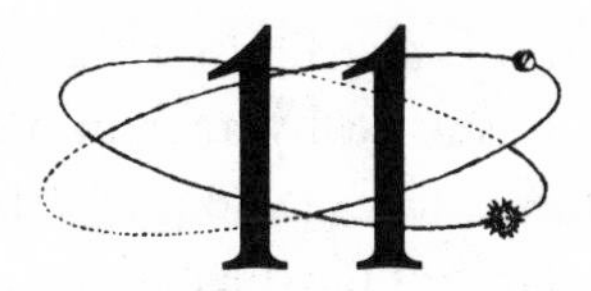

Celestial Disharmonies

•

Our souls, whose faculties can comprehend
The wondrous Architecture of the world:
And measure every wand'ring planet's course,
Still climbing after knowledge infinite.

CHRISTOPHER MARLOWE (1564–1593),
Conquests of Tamburlaine

PIERRE-SIMON de Laplace gave the field of celestial mechanics its name. He was also a member of the elite group of mathematicians and astronomers who founded the Bureau des Longitudes in Paris on June 25, 1795. An expression of the passion for scientific order and rationality that coexisted with the disorder of revolutionary France, this act produced an institution dedicated to precise measurement and timekeeping. Its duty was to monitor the well-regulated clockwork of the solar system.

Laplace labored on the problems of predicting planetary and lunar motions with sufficient accuracy to provide reliable tables for navigation. Today, Jacques Laskar finds himself immersed in many of the same issues that preoccupied his illustrious predecessor at the Bureau des Longitudes two centuries ago. He, too, attacks questions of the solar system's stability and long-term future, but his fresh insights, molded and refined in the crucible of computation, reveal a solar system inescapably teetering on the edge of chaos.

The journey toward this startling state of affairs began with the compelling regularities observed by ancient astronomers in the motions of the night sky. Kepler, Galileo, and Newton went beyond description to true prediction and explanation, extracting from flawed, confusing data and the uncertainty and ambiguity of experience the remarkably simple, fundamental laws that govern the universe. They invented an ideal mathematical world that could be held up as a mirror to nature and could be controlled and manipulated as never before. In this triumph of mind over matter, the near perfection of the solar system showed the way, furnishing the cleanest data and the first definitive tests of the new theories.

By means of these mathematical surrogates for the real world, scientists and mathematicians pursued the achievable: solving equa-

Jacques Laskar. (Courtesy of Jacques Laskar.)

tions, analyzing solutions, deriving simple formulas. Celestial mechanics and other scientific pursuits were a record of one triumph after another. In the midst of these successes, scientists and mathematicians were slow to recognize the chaotic phenomena that lurked just beyond their calculations. But the unruliness of everyday experience should have provided a clue that there was much more to Newton's laws than the perfection suggested by the few examples of differential equations that could be integrated to a simple formula. It was Henri Poincaré who, by turning to the general and the qualitative, became the first to glimpse the tangled possibilities that lie hidden in Newton's majestic clockwork.

Laskar's journey into chaos started quite innocently nearly a decade ago with the delicate problem of accurately determining the changes that had occurred in Earth's orbit over the last several million years. Such data would enable researchers to evaluate possible links between subtle shifts in Earth's orbit and variations in its climate. But the traditional, cumbersome machinery of celestial mechanics proved unequal to the task. Laskar quickly realized that it was pointless to try to obtain approximate solutions in the form of lengthy algebraic expressions of the type derived by Laplace and Le Verrier, whose calculations had led to the discovery of Neptune (see Chapter 5). As Poincaré so pointedly demonstrated a century ago, solutions expressed in these forms generally fail to converge to a specific answer. This gradual drift away from certainty becomes problematic when the calculations must be accurate enough to reproduce orbital fluctuations over millions of years.

Laskar decided to split the problem into two parts. He abandoned any pretense of trying to predict the precise future positions of the planets on time scales spanning millions of years or more. He realized that although he could come up with no meaningful timetable for the planets, he might be able to provide a reasonable map of their paths.

Instead of using the full equations of motion, Laskar focused on a special formulation that spotlights gradual, but cumulative, changes in an orbit's shape and orientation. He worked with equations that smooth out the recurring wiggles and wobbles in planetary orbits, leaving only the long-term trends. It's like the strategy that stock market analysts employ to extract a modicum of meaning from the erratic ups and downs of daily share prices. After defining the Dow

Jones industrial average, which summarizes the behavior of 30 different stocks and serves as a compact surrogate for the entire stock market's performance, they look at averages over 30 days, each day dropping the oldest value and adding in the most recent one.

By applying a similar strategy to celestial curves, Laskar could isolate those parts of a planet's motion that correspond to lasting changes in key characteristics of its orbit. To accomplish this prestidigitation, he used computer algebra—an elaborate system of computing with symbols rather than digits—to painstakingly construct a mathematical expression that eventually stretched to 150,000 algebraic terms. This monstrosity allowed him, in effect, to spread out the masses of the planets evenly along their orbits. Then he could use a computer to integrate these "averaged" differential equations to track the evolution of planetary orbits.

Laskar soon found himself expanding the purview of his work from teasing out Earth's wobbles in aid of climate studies to questioning the entire solar system's stability. To begin his search for some hint of the solar system's ultimate fate, Laskar defined an initial state, specifying the positions, velocities, and masses of all the planets except Pluto at a given moment. Then he let a supercomputer grind through the calculations to map out the solar system's evolution, tracking planetary paths for 200 million years in 500-year increments.

To test for chaos, he repeated the computation with a slightly different starting point and compared the two results. If the orbits were approximately regular (or quasi-periodic), then the two computed trajectories of the solar system through phase space (the abstract space in which one tracks not only position but also velocity or momentum) would stay close together, separating no faster than at a rate proportional to the time elapsed. In contrast, chaos would be evident if their separation dramatically doubled, then redoubled, and so on, with the passage of equal amounts of time. Indeed, the more chaotic the system, the less time it takes for the separation between two otherwise identical bodies in phase space to double. With such an amplification of initially minute differences, prediction of a system's past or future becomes extremely sensitive to its present state and to the precision with which that state can be measured. Hence, we can never accurately forecast its future or recover its past.

That's how Laskar unveiled chaos in the apparent evolution of the solar system. According to his equations, the separation between the paths of two solar systems starting at slightly different positions and velocities in phase space would double every 3.5 million years. A difference in starting point of just one part in 10 billion—an error as small as 100 meters in measuring Earth's position at a given moment—would mount so quickly that it would be impossible to specify where Earth would be in its orbit 100 million years later.

Here was further evidence of chaos among the planets. As described in Chapter 10, Jack Wisdom and Gerald Sussman had determined that Pluto's motion has chaotic components. Now Laskar's computations clearly demonstrated that under given conditions, a body identical to Earth but at a minutely different position in its orbit could easily end up on a very different trajectory. Laskar's result, however, didn't necessarily mean that Earth is likely to wander from its usual path in the next 10 million years or so, perhaps to end up on a collision course with Mars or Venus. In a curiously ambiguous fashion, one can calculate that Earth will *probably* stay roughly the same distance from the sun for the next 100 million years, but one can't be *absolutely* sure that this will happen.

Laskar emphasized this point in a paper published in the journal *Nature* in 1989: "This does not mean that after such a short timespan (geologically speaking) we will see catastrophic events such as a crossing of the orbits of Venus and Earth; but the traditional tools of quantitative celestial mechanics (numerical integrations or analytical theories), which aim at unique solutions from given initial conditions, will fail to predict such events." In other words, the predictability of the orbits of the inner planets, including Earth, declines sharply within a few tens of millions of years. Laskar stated his conclusion cautiously: "The computation shows that the solutions of our differential system, with the initial conditions of the Solar System, are chaotic. This differential system is a close approximation to the real Solar System. We can conclude that the motion of the Solar System, and in particular the inner Solar System, is close to being chaotic, but the exact meaning of 'close' is still difficult to evaluate."

As the continual stream of polite sniping that inevitably arises in the scientific world attested, many at first found these results unconvincing. The methods were too new and untested. There were questions of reliability and suspicions that the results represented artifacts of computation rather than real phenomena. Even the very notion of chaos, and the means used to identify its presence, met with withering scrutiny from mathematicians skeptical that astronomers could know what they were doing mathematically. Similar criticisms had greeted the earlier work of Wisdom and Sussman and of others who relied on complicated mathematical stratagems and intensive computation for their results.

Laskar tried to counter some of these criticisms in a lengthy paper published in *Icarus* the following year. In particular, he addressed the question of what it is about the planets and their particular arrangements and interactions that produces this intrinsic uncertainty. Laskar was able to identify specific terms in his averaged equations that pointed to two, previously unidentified, complex but subtle gravitational interactions between certain planetary motions as the main source of this surprising chaotic behavior. One of these interactions, or resonances, involves Mars and Earth; the other involves Mercury, Venus, and Jupiter.

Here was the fatal defect: Resonances destroy predictability. They amplify small effects into potentially significant forces and skew any effort to compute long-term trajectories in their vicinity. What's fascinating is that in dynamical systems like the solar system, in which the sun has by far the largest influence and all other bodies perturb the sun's predominant effect only slightly, resonances lead to chaos and strongly influence where it resides. Chaos doesn't occur just anywhere in the solar system. Laskar concluded: "The large variations in the fundamental frequencies of the inner Solar System in the course of its history implies a new vision in the study of several long term phenomena in the history of the Solar System. . . . The picture of the dynamics of the Solar System which is given here is very different from the regular quasiperiodic motion described by Laplace and Lagrange 200 years ago. It is much more complicated."

Because Laskar used averaged equations, one possible test of his conclusions involved integrating the full equations of motion. As it happened, Scott Tremaine of the Canadian Institute for Theoretical Astrophysics at the University of Toronto and his collaborators had been refining their own methods of computing planetary orbits over long time intervals. In earlier studies, Tremaine, Martin Duncan, and

Thomas Quinn had used an alternative mathematical stratagem to highlight particularly significant planetary interactions in their search for potentially stable orbits between planets. They hoped to identify regions of space that might harbor hitherto undetected debris left over from the formation of the solar system.

Using a computer to track the behavior of 300 test particles—massless points that respond to the gravitational tugs of simulated planets but exert no influence themselves—in a dynamical system that behaved approximately like the real solar system, they discovered that many orbits lying between Uranus and Neptune become chaotic. This supported the notion that Pluto may have entered its present unusual orbit via a chaotic trajectory. Of particles that started out between Uranus and Neptune, roughly half their orbits became

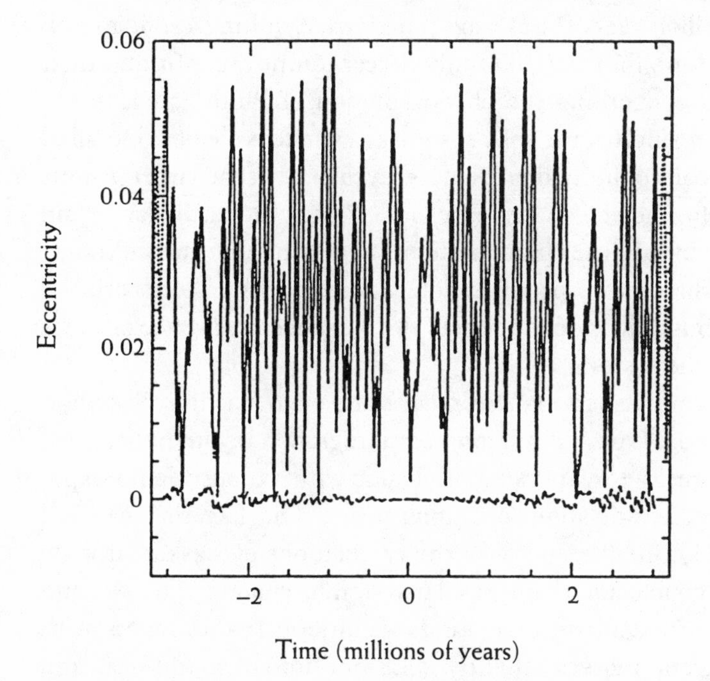

The eccentricity of Earth's orbit as computed over a six-million-year time span centered at the present. The solid line representing the solution obtained by Quinn, Tremaine, and Duncan closely follows the dotted line representing Laskar's result. The dashed line at the bottom illustrates the difference between the two solutions. (From Laskar et al., *Icarus*. New York: Academic Press, 1992.)

chaotic enough over five billion years to be ejected from the solar system.

Working independently of Laskar, the same team also used computers to solve directly the equations of motion for the solar system's planets. Taking particular care to make their model as physically accurate as possible by including various effects, including those produced by general relativity, they tracked the evolution of these orbits over a six-million-year period—three million years into the future and three million years into the past. Although this computation couldn't directly confirm the chaotic behavior Laskar had discerned, it picked up the same resonance he had noted between Mars and Earth.

At this stage, Wisdom and Sussman entered the picture. In 1992, using their new, custom-built Supercomputer Toolkit, they traced the evolution of the entire solar system over a period of 36 billion days, or nearly 100 million years. These computations, requiring a solid month of computer time for each 100-million-year journey, confirmed their earlier results concerning the chaotic motion of Pluto and Laskar's more general result that the solar system as a whole is chaotic. Detailed looks at the orbits of individual planets revealed that the larger planets independently showed signs of chaos. Pluto, in particular, again showed large instabilities in its motion, suggesting that the unknown physical mechanism responsible for generating its chaotic orbit is extremely robust and independent of the presence or absence of chaos in the rest of the system.

In a paper published in *Science* that same year, Wisdom and Sussman concluded: "Our 100-million-year integration of the entire solar system indicates the solar system is chaotic with a timescale for exponential divergence of about 4 million years. The fact that we find similar behavior in all respects to the calculations of Laskar strongly supports the conclusion that the solar system is chaotic. That we and Lasker have carried out different kinds of numerical experiments, with slightly different masses, slightly different initial conditions, and slightly different physics, shows that the chaotic character of the solar system is not sensitively dependent on the precise model or numerical methods."

The agreement among these vastly different approaches to computing the solar system's dynamical evolution provided strong indirect support for the presence of chaos. Moreover, there was a

The retired Digital Orrery rests atop its successor, the Supercomputer Toolkit. (Courtesy of Gerald Jay Sussman.)

possible explanation for chaos in the resonances that Laskar had detected in his equations. Evidently, chaotic dynamics plays some kind of role in the solar system and may have something to say about the system's long-term stability. But, as Wisdom and Sussman warned, "we will not fully understand the consequences of the observed chaotic evolution until we clearly understand the dynamical mechanisms responsible for it."

Because Laskar, Tremaine, and Wisdom employed very different methods, the close agreement of their results also furnished an important independent check on the techniques used. Indeed, the availability of reliable techniques for computing reasonably realistic planetary orbits over moderately long time periods has generated a new field of study. Researchers can now, in effect, create fictitious solar systems to gain a sense of the variability allowed within a sun-centered planetary system.

In 1991, Gerald Quinlan, working with Tremaine, started a program of following the evolution of orbits in ersatz solar systems consisting of four planets similar in mass, position, and orbit to Jupiter, Saturn, Uranus, and Neptune. In simulations of more than 50 such fictitious, randomly adjusted solar systems, a majority showed at least mild symptoms of chaotic behavior. In one instance, shifting Saturn's position even slightly to make its orbit a little wider could drive the entire system into chaos. Summarizing this preliminary work, Quinlan observed, "Chaos appears to be a common feature of planetary systems, even when the planets are started on orbits that are well separated. Because so many of these [randomly] modified solar systems . . . were chaotic, it's not surprising that our real solar system is chaotic. It would be surprising if it weren't."

In general, the resulting computed orbits looked somewhat more erratic than they are in the real solar system; at the same time, however, nothing terrible ever happened to the orbits. They didn't cross, and none of the planets was ejected from the system over a period of a few million years. In other words, constrained chaos showed up far more often than chaos of the catastrophic variety.

Such computer experiments may have something to say about the extent to which stability requirements force the solar system to look the way it does. Is the particular distribution of planets in our solar system just one of many possible stable arrangements? Or is it the one

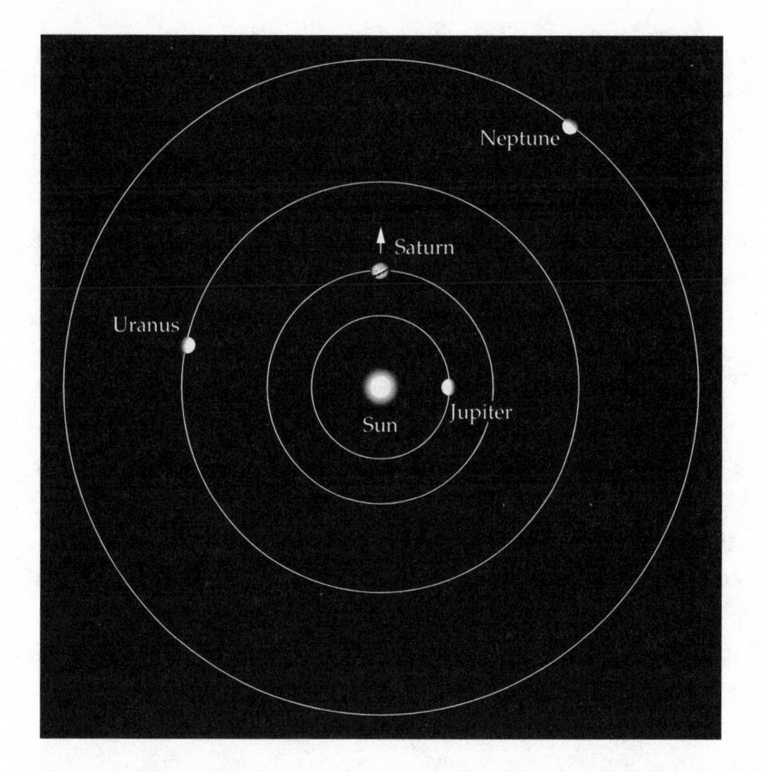

In a fictitious solar system consisting of four planets (twins of Jupiter, Saturn, Uranus, and Neptune), slightly shifting the orbit of Saturn so that the planet's mean distance from the sun is somewhat larger than in the real solar system produces hints of chaotic motion. Such a change, though small, pushes the system much closer to the 5:2 resonance between the mean motions of Jupiter and Saturn.

that endures because it is stable? Such questions remain largely unresolved.

Opinions also differ on the related question of whether the solar system would have room for an extra planet without its apparent equanimity being disturbed. Some researchers, including Peter Goldreich of Caltech, suspect that any additional material would be at risk of ejection from the solar system. This sort of ejection mechanism may have led to precisely the configuration of planets and asteroids we see today, in which the survivors represent largely the most massive chunks of material. Such large objects are much harder to perturb or eject than miniscule fragments bouncing about in the wake of their massive compatriots.

Indeed, some planetary scientists speculate that several billion years ago the solar system may have had a number of extra planets, perhaps the size of the moon or Mars, that were thrown out at some point. Computer simulations show that the orbits of bodies the size of asteroids can suddenly change their eccentricity, causing collisions or ejections. It's possible that the same thing can happen to interloper planets, given sufficient time. In addition, a proto-Jupiter gradually shifting to an orbit nearer the sun could readily sweep up material left over from the swarm of planetesimals probably present at the formation of the solar system, leaving a broad, relatively clear zone in its wake.

One astronomer has suggested that the solar system once contained thousands of Plutos as a consequence of its formation. In such a case, the ejection of a large proportion of these bodies would be virtually inevitable. With no way to view such departed bodies directly, however, astronomers testing this scenario must rely on evidence supplied by computer simulations of dynamical conditions early in the solar system's history.

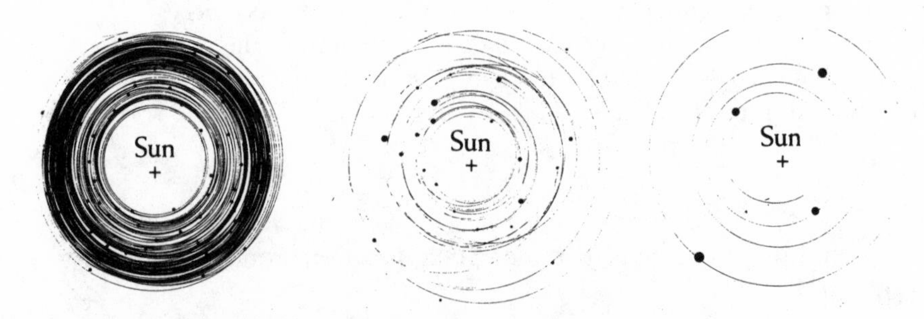

These three drawings show the results of a computer simulation of the formation of the inner planets. *Left:* The simulation begins with 100 planetesimals. *Middle:* After 30 million years, these planetesimals have coalesced into 22 protoplanets. *Right:* This final view is for an elapsed time of 441 million years, but the formation of the inner planets was essentially complete after 150 million years. (Adapted from G. W. Wetherill. Reprinted with permission from William J. Kaufmann, *Universe,* 3d ed. New York: W. H. Freeman, p. 142.)

The role of chaos in the formation of the solar system remains a controversial issue. There's no clear consensus on whether the solar system settled down out of the primordial cloud of gas and dust into its present configuration—with well-spaced planets having near-circular orbits lying in roughly the same plane—in the first few million years of its existence, or whether it gradually evolved to its present state over the past five billion years or so. It's certainly possible that chaos played only a limited role in the dynamical history of the solar system, merely sweeping out the areas between the major planets and establishing the intricate dynamics of the asteroid belt. At this stage, no one knows whether chaos was also important in clearing out the regions between Venus and Earth and between Earth and Mars.

Nonetheless, researchers have found evidence of some sort of chaotic behavior in practically every long-term integration yet undertaken. Such results suggest that chaos truly does pervade the solar system. Even the solar system's general character—nine widely spaced major planets with only a little debris between them—hints at the role chaotic dynamics may have played in its formation.

At the same time, efforts to settle the question of the solar system's stability face a serious, perhaps insurmountable, obstacle. As Scott Tremaine has remarked, "In some sense, you end up having to deal with probabilities. You can never rule anything out completely. Even if a system is well-behaved, there's always a small chance of its wandering by some narrow path to just about any configuration." In other words, with a mathematical model that automatically incorporates chaotic behavior, there's no way to prove with absolute certainty that something can't ever happen. This built-in constraint reflects something in the nature of the mathematics involved and is entirely independent of the customary uncertainties associated with the quality of measurements, the formulation and applicability of physical "laws," and the frequently tenuous connections between complicated reality and simplistic mathematical models.

Another question concerns the meaning of the exponential divergence of nearby orbits—as measured by a number known as the Lyapunov exponent or Lyapunov time—in a chaotic system. Laskar's solar system calculation produced a Lyapunov time of roughly 5 million years; Wisdom and Sussman's work produced a value of 25 million years for Pluto's orbit and 4 million years for the solar system.

Although widely used as an indicator of chaos, the value of this exponent doesn't necessarily reflect the time scale over which a drastic change in a single orbit or an entire system could occur.

Researchers note that defining the existence of chaos in terms of the exponential rate at which nearby trajectories diverge provides just one technical fact about the system. Whether that behavior has any qualitative implications over the lifetime of the solar system is a much more intriguing, yet largely unanswered, question.

All this remains worrisome. Sussman and Wisdom have remarked: "The fact that almost all long-term integrations of the solar system give exponential divergence of trajectories with a time scale in the range of 3 to 30 million years in physically quite different models is very striking and unsettling. Another unsettling feature of the chaotic behavior we observe in long-term planetary integrations is that nothing dramatic happens. This is compounded by the fact that [despite the efforts of Laskar and others] in no case have we unambiguously identified the mechanism responsible for the chaotic behavior."

Indeed, the solar system's survival in roughly its present form for billions of years—many times longer than its computed Lyapunov exponent would suggest—clearly indicates that the underlying dynamical theory requires more work. Chaos, which implies erratic behavior, seems somehow incompatible with the planets going around the sun for a long time without doing anything crazy. Even the asteroids—flotsam on a dynamical sea—wander on a leash.

The clue to what may be going on is the fact that there's sometimes a limit within a physical system on how far two trajectories in phase space can diverge. As an analogy, imagine two nearby points on the surface of a sphere. As they move apart, their separation initially increases; over time, however, their wanderings take them all over the sphere's surface, and occasionally they are near each other once again. Moreover, the two points can never be farther apart than half the sphere's circumference. So if the distance between them starts out doubling and redoubling, eventually there will be no more room in which the divergence can continue. Similarly, trajectories in phase space often have only a finite realm within which to wander.

In 1992, Myron Lecar, Fred Franklin, and Marc Murison, researchers at the Harvard-Smithsonian Center for Astrophysics, took one step toward solving the problem of why the presence of chaos and

Myron Lecar *(top)*, Fred Franklin *(middle)*, and Marc Murison *(bottom)* at the Harvard-Smithsonian Center for Astrophysics. (Courtesy of Myron Lecar.)

its attendant unpredictability are compatible with the solar system's equanimity over billions of years. Using computer simulations, they studied more than a thousand examples of three types of orbits. In one case, they computed sample orbits traced out by an asteroid gravitationally influenced by only the sun and Jupiter. In another scenario, they looked at orbits of hypothetical asteroids stationed between Jupiter and Saturn. The third situation involved tracking the orbit of a tiny body initially circling the smaller of two stars in a binary system.

They computed a Lyapunov time—the characteristic time over which an orbit is said to remain predictable—for each situation, providing an estimate of how far into the future its chaotic behavior could be predicted. The researchers then compared the computed Lyapunov time with the time it takes orbits to change sufficiently for an orbiting body to cross a planet's path or escape the system. They discovered that, despite the great differences among the three types of orbits considered, all showed approximately the same numerical relationship between the Lyapunov time and the much longer period after which an orbit is likely to make a sudden, drastic transition to a new eccentricity or orientation. According to this relationship, the transition time is proportional to the Lyapunov time raised to the 1.8 power.

For the solar system, the results imply that no catastrophe is likely to occur for at least a trillion years. This is a long time, even by astronomical standards. Indeed, well before such a period elapses, the solar system is much more likely to be engulfed by an explosively swelling sun shifting from the consumption of one nuclear fuel to another toward the end of its life.

Is there a chance that the solar system might get into trouble much sooner? Yes, but the probability is very small—though not zero.

Lecar likens the process underlying these results to the wandering of a tiny speck initially confined to the tight space between the two innermost members of a set of snugly nested, hollow eggs. The relatively small Lyapunov time characteristic of such a system reflects the limited scope of the speck's movements in this ellipsoidal gap. However, the speck can also drift by way of secret passages to an adjacent gap between the next two, larger eggs, and so on, until it reaches the last egg, when it becomes free to roam in a much vaster space. The time it takes the speck to "diffuse" by random steps to this wider space corresponds to what Lecar calls the "transition time." Nothing much happens until the speck has diffused all the way through, and only when it's no longer trapped can it drastically change its dynamical behavior.

If such a mechanism is at work in a dynamical system, then statistics plays a key role. In a few instances, the speck may find the right passages through its dynamical eggshells and make it all the way to the outside in a short time. Drastic changes could then occur much sooner than expected. In other cases, the speck may remain trapped for an extremely long time. This suggests that if one were to integrate realistic equations of motion for the solar system, taking it 4.5 billion years into the future (much less than its predicted transition time of a trillion years or so), the probability is high that Earth's orbit would look very much as it does now. But if that computation were done, say, a billion times, it's possible that one of the integrations would show a dramatic change in the size, shape, or inclination of Earth's orbit.

However, Lecar and his collaborators looked at only three special cases, involving gravitational interactions much less complicated and among fewer bodies than those in the solar system. What their analysis implies for real planets and other chaotic systems remains unclear. Some scientists suspect that further studies will show that a single, universal way of characterizing exactly what's going on is not likely to be found. In many cases, the details of what happens depend quite strongly on the actual arrangement of the bodies in the system under consideration. Also missing is a convincing explanation of why the exponent in the cases studied happens to be 1.8.

Martin Duncan of Queen's University in Canada has been studying the long-term dynamical behavior of hypothetical asteroid orbits beyond Neptune. This particular region of the solar system encompasses what's known as the Kuiper belt, which may serve as a vast repository of comets. These comets may spend billions of years in storage, inconspicuously orbiting the sun until their orbits suddenly become eccentric enough to swing them out of the belt and across Neptune's orbit, there to be flung by Neptune into the sun's glare.

Duncan's simulations, which cover up to a billion years, confirm that these objects have Lyapunov time scales similar to those computed by Lecar and his group. And because Duncan can track the asteroids' motions much farther into the future, he can also confirm directly that they spend hundreds or thousands of times longer in normal orbits than the Lyapunov time by itself would indicate. These objects take a long time to leave the Kuiper belt, although it's impossible to predict precisely the position of any one of them much beyond the time given by the Lyapunov exponent.

This apparent stability in the face of unpredictability remains a deep puzzle. Pluto, for example, has apparently remained stable for a long time, far longer than the limit of predictability of its orbit. And the inner planets and asteroids, though they seem even more strongly chaotic than Pluto, apparently haven't undergone any dramatic changes in orbit.

So chaos and instability aren't necessarily synonymous. For example, the 3:2 resonance between Pluto's orbit of the sun and Neptune's period may actually protect Pluto from ejection out of the solar system. On a larger scale, the solar system itself may have within it some subtle interaction that somehow keeps it trapped in its present configuration.

Thus, it's possible that chaotic motion in the solar system, which precludes accurate predictions far into the future, may be limited such that orbit crossings and other catastrophes are averted. Whether the solar system will continue on roughly its present course without major alteration remains an open question, but its past history certainly suggests that it probably remains stable for geologically significant periods.

This curious and subtle situation regarding stability was foreshadowed in the mathematical work conducted by Andrei N. Kolmogorov in the 1950s and by several mathematicians who followed in his footsteps. The combined work of Kolmogorov, his student Vladimir I. Arnol'd, and Jürgen Moser provided precisely defined mathematical criteria for determining whether perturbations can push a dynamical system into instability. From Poincaré's work and that of his successors, mathematicians already had some sense that the phase spaces of systems containing three or more bodies are characterized by an intricate interweaving of regular and chaotic regions. The question that remained was whether the feeble perturbations of the planets, compared with the sun's overwhelming effect, are sufficient to lead to true instability.

Through what is now known as KAM theory (the acronym is derived from the surnames of the theory's three principal authors), mathematicians can demonstrate that motion in any dynamical system remains for the most part regular, or quasi-periodic, if perturbations stay small. In other words, a large set of initial conditions leads to quasi-periodic rather than chaotic orbits if the masses, eccentricities, and inclinations of the planets are sufficiently tiny. In addition, the length of time that the planets take to complete a circuit of the sun may not correspond to a simple ratio, such as 1:2, 1:3, or 2:3.

The orbits that permit regularity in the face of small perturbations are precisely those for which no resonance occurs. Under such conditions, the planets could not readily sail off on their own, and the system would be stable. However, because resonating orbits can lie close to one another, it's entirely possible for an imperceptible shift in a starting point to turn a regular orbit into an unstable one. Indeed, Arnol'd has noted that the initial conditions leading to regular or chaotic behavior alternate in a very intimate manner. Thus, even if the motion of a celestial body is regular, an arbitrarily small perturbation of its initial state is sufficient to make it chaotic.

Arnol'd has expressed his own conclusion regarding the solar system's fate in light on these mathematically defined constraints: "Fortunately, however, the rate of development of these chaotic perturbations is extremely small, so the time for which chaos manifests itself under a sufficiently small perturbation of the initial state is large in comparison with the time of existence of the solar system. So for the next billion years the main part of the solar system will hardly change and the 'clock mechanism' described by Newton will continue in good working order."

Laskar, however, has noted that the solar system—particularly the inner planets—resides in a suburban neighborhood of phase space, far from the central business district where planetary masses, eccentricities, and inclinations are small and meet the strictures of KAM theory favoring stability. Nonetheless, the purely mathematical results derived by Poincaré and KAM theory provide a good framework for the study of the solar system's stability. What mathematics does not give is the precise location of the actual solar system within this generic picture. Laskar remarks that "in fact, the Solar System is much too complicated, with far too many parameters, to obtain mathematical results on its stability at present." That can be determined only indirectly through intensive computation. It's a branch of celestial

HARMONICIS LIB. V. 207

mnia (infinita in potentiâ) permeantes actu: id quod aliter à me non potuit exprimi, quam per continuam seriem Notarum intermedia- CAP. VI

rum. Venus ferè manet in unisono non æquans tensionis amplitudine vel minimum ex concinnis intervallis.

Atqui signatura duarum in communi Systemate Clavium, & formatio scelecti Octavæ, per comprehensionem certi intervalli concinni, est rudimentum quoddam distinctionis Tonorum seu Modorum: sunt ergò Modi Musici inter Planetas dispertiti. Scio equidem, ad formationem & definitionem distinctorum Modorum requiri plura, quæ cantus humani, quippe intervallati, sunt propria: itaque voce quodammodò sum usus.

Liberum autem erit Harmonistæ, sententiam depromere suam: quem quisque planeta Modum exprimat propiùs, extremis hic ipsi assignatis. Ego Saturno darem ex usitatis Septimum vel Octavum, quia si radicalem ejus clavem ponas *G*, perihelius motus ascendit ad ♮: Jovi Primum vel Secundum; quia aphelio ejus motu ad *G* accommodato, perihelius ad *b* pervenit; Marti Quintum vel Sextum; non eò tantùm, quia ferè Diapente assequitur, quod intervallum commune est omnibus modis: sed ideò potissimùm, quia redactus cum cæteris ad commune systema, perihelio motu *c* assequitur, aphelio ad *f* alludit: quæ radix est Toni seu Modi Quinti vel Sexti: Telluri darem Tertium vel Quartum: quia intra semitonium ejus motus vertuntur; & verò primum illorum Tonorum intervallum est semitonium; Mercurio verò ob amplitudinem intervalli, promiscuè omnes Modi vel Toni convenient: Veneri ob angustiam intervalli, planè nullus; at ob commune Systema, etiam Tertius & Quartus; quia ipsa respectu cæterorum obtinet *e*.

Tell^s canit MI FA MI ut vel ex syllaba conjicias, in hoc nostro domicilio MI seriam & FAmem obtinere.

CAPVT VII.

Harmonias universales omnium sex Planetarum, veluti communia Contrapuncta, quadriformia dari.

NVnc opus, Vranie, sonitu majore: dum per scalam Harmonicam cœlestium motuum, ad altiora conscendo; quâ genuinus

As illustrated in this page from his *Harmonices mundi* (Harmonies of the world), Johannes Kepler tried to link musical harmonies with the orbits of the planets. But his vision of a harmonious, orderly solar system eventually foundered on the shoals of chaos. (Library of Congress.)

mechanics that is nearly an experimental science, with mathematical proofs lagging far behind.

Far from finding the definitive proof of stability that eluded Newton, Laplace, and Poincaré, today's theorists have raised the possibility that there may not be one. Uncertainty is the most certain thing about planetary orbits; the solar system doesn't really run like a clock. Sensitive dependence on initial conditions rules, and chaos lurks everywhere. Any assertion concerning the behavior of planetary orbits can only be probabilistic and must be investigated by computing many solutions with different initial conditions to obtain a range of possibilities.

More than 300 years after publication of the *Principia,* the full implications of Newton's deceptively simple law of gravity, with its surprisingly complicated consequences, still elude us. One has only to look at the strangeness of a chaotically tumbling satellite like Hyperion or at the intrinsic difficulties of calculating the moon's itinerary or delving into the solar system's origins to sense the dynamical mysteries that confront us.

Studies of chaotic dynamics now provide a revealing link between the idealized, pristine realm of abstract physical laws and the disordered, complex world in which we actually live. These insights help to define more clearly the limits of what we can accomplish, while suggesting new landscapes worthy of exploration. As Jack Wisdom concluded in his 1986 lecture delivered in acceptance of the prestigious Urey Prize for his asteroid work, "In the early part of this century, classical mechanics was eclipsed by quantum mechanics, and rightly so. Quantum mechanics provides a better description of the world. However, the relevance of classical mechanics is once again beginning to be appreciated. For the most part, the world of our everyday lives is classical, and classical mechanics is not at all simple. Newton could not have dreamt of the beauty and complexity of the mechanics that he brought forth. The final state of Hyperion is completely unpredictable. Apparently, even in a classical world God does, after all, play dice."

For Jack Wisdom, Jacques Laskar, and others, the search for chaos in the solar system has continued. Toward the end of 1992, Wisdom became intrigued with the notion of transforming his many plots and tables into a movie showing the evolution of planetary orbits. He wanted to see more clearly and vividly what the chaos that he had uncovered really meant.

Wisdom's starting point was a scale diagram of the planetary orbits as seen from high above the solar system's orbital plane. Because there was no stretching of the orbits to accentuate their eccentricities, nearly all the orbits on this scale looked approximately circular, and the orbits of the inner planets were barely distinguishable from the sun in the middle. With orbits drawn as nested rings, the diagram resembled any of the typical illustrations of the solar system found in countless textbooks.

But planetary orbits change over time. Wisdom chose to compress 60,000 years of solar system history into each second of his computer simulation. Thus, each second of the resulting movie represented a time span significantly longer than the period over which astronomers—ancient and modern—have been observing the heavens. And the results were dramatic.

The rings representing the orbits of Jupiter, Saturn, Uranus, Neptune, and Pluto seemed to be in continuous motion, restlessly jiggling to jittery, complex rhythms. Though the rings never quite crashed together, they seemed to knock one another around. For instance, as the orbits of Pluto and Neptune bounced back and forth against each other, there were times when Pluto's orbit was entirely outside that of Neptune. At other times, the two orbits crossed. Uranus acted as if it were being erratically kicked around by its neighbors. Jupiter plainly had a strong influence on Saturn's path.

A closeup of the inner planets showed a similar restlessness. Earth certainly wobbled and wandered in its orbit, but its mildly erratic motion paled in comparison with the wild gyrations and vibrations of Mars. It was also easy to see the rotation of Mercury's elliptical orbit, along with a host of more subtle changes. The entire solar system seemed to vibrate with a mesmerizing energy—an energy that freely sloshed back and forth among the various planets. What Johannes Kepler would have made of such startling complexity no one can guess.

When Wisdom showed his videotape at a meeting early in 1993, he reflected, "I think if you were given this movie and then asked to prove the stability of the solar system, you would right away say: Hummmm. I'm not so convinced the solar system *is* stable."

Curiously, Wisdom's movie showed no collisions. Despite their erratic vibrations, the rings stayed apart from one another. No one is really sure why. "Chaos doesn't necessarily mean catastrophe," Wisdom commented.

One of the most dramatic consequences of this chaotic evolution of planetary orbits is its effect on the angle of a planet's spin axis. Earth's axis, which passes through the planet's north and south poles, is tilted 23.5 degrees away from a line at right angles to Earth's orbital plane. This modest tilt is responsible for the cycle of the seasons and has apparently remained close to this value for millennia. In contrast, new calculations by Wisdom and others show that the tilt of Mars can change drastically in just a few million years. These oscillations can be so extreme that the planet's spin axis sometimes dips enough to bring more sunlight to polar regions than to equatorial zones.

In Earth's case, the moon acts as a stabilizing influence. The moon's bulk forces the rotation, or precession, of Earth's axis at a sufficiently rapid rate to forestall wildly erratic variations in Earth's tilt. Although one complete rotation, as observed in the precession of the equinoxes, requires 26,000 years, that's rapid enough to keep it out of the range of chaos-triggering resonances with other motions in the solar system. Thus the moon may play a crucial role in regulating Earth's climate, stabilizing it enough to provide a comparatively equable setting for the evolution of life. Indeed, the presence of a moon-sized satellite may be a necessary condition for finding Earth-sized planets with Earthlike climates in orbits around neighboring stars.

Mars, on the other hand, has no such satellite to shift its motion away from resonances. Years before Wisdom's work, researchers had already become aware that a resonance could occur between the precession rate of the spin axis of Mars and one of the characteristic frequencies of the solar system's motion as a whole. They had speculated that intense volcanic activity and other geological processes could alter the planet's spin sufficiently to bring these rates exactly into resonance and thereby cause a large shift in its tilt angle.

Wisdom's analysis showed that such changes are larger than expected and essentially unpredictable. In an independent study, Jacques Laskar came to roughly the same conclusion. Geological processes are not the only possible causes of drastic changes in Mars's spin axis. The tilt of the planet's axis, together with its chaotic orbit, produces large, irregular oscillations that cause dips as large as 50 or 60 degrees in the tilt angle. For planetary scientists studying the climate of Mars and the evolution of its surface, these new results drastically alter the kind of information they must put into their computer models to capture Mars's past and predict its future.

Laskar and his colleagues have gone a step further by suggesting that the tilts of all the inner planets could have evolved chaotically at various times in the past. Earth itself may even enter such a chaotic zone when the distance between Earth and a slowly departing moon reaches 68 Earth radii in a few billion years. (The present-day distance is about 60 Earth radii.) Given that variations in tilt angle as small as 2 degrees can trigger ice ages, the forecast for Earth when its axis shifts to an angle of nearly 60 degrees would certainly be bleak.

Long held up as a model of perfection and the symbol of a predictable mechanical universe, the solar system no longer conforms to the image of a precision machine. Chaos and uncertainty have stealthily invaded the clockwork.

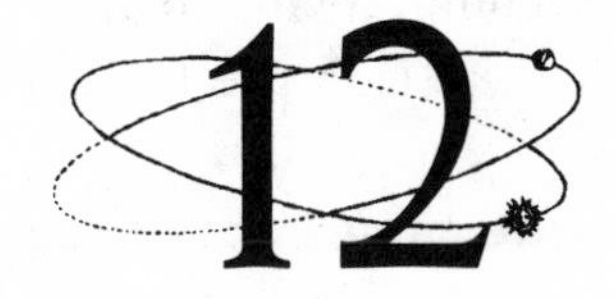

12 Machinery of Wonder

•

Atoms or systems into ruins hurl'd,
And now a bubble burst, and now a World.

ALEXANDER POPE (1688–1744),
An Essay on Man

MATHEMATICS PLAYS an enigmatically central role in the search for order, predictability, and pattern in the physical world. It promises what amounts to absolute certainty. In their everyday pursuits, people expect that the same set of numbers will always add up to the same answer. The sum of the interior angles of a triangle is 180 degrees. There are precisely 17 fundamentally different ways in which patterns can be repeated endlessly in two dimensions. Shopkeepers, surveyors, and architects rely on such certainties to go about their daily business.

Armed with an extensive array of these abstract notions, scientists can pluck from their mathematical tool kits the hammers and screwdrivers they need to bang out predictions and take apart problems. These triumphs of mathematical reasoning, particularly combined with the effort to understand heavenly motions and uncover universal harmonies, go back thousands of years. They testify to the astonishingly enduring power of mathematics in organizing human thought and provide a counterpoint to the turmoil—political and otherwise—that has characterized the human condition throughout history.

Even in the turbulent waning days of the Roman Empire, for example, mathematical scholarship endured. Invaders bent on pillage and conquest continually threatened Roman settlements, venturing as far as Rome itself. The Vandals established a kingdom in Spain, only to fall to the Visigoths, who had swept across Greece, plundered Athens, and invaded Italy. The Franks, Alemanni, Ostrogoths, and Huns all claimed their pieces of the crumbling empire. Finally, in 476, German occupiers elected one of their own generals, Odoacer, to reign as king of Italy, effectively deposing the ruling emperor, Romulus Augustus.

The early part of the fifth century also saw acrimonious theological disputes that contributed to the sense of disorder and helped split apart the western and eastern parts of the empire. The Roman emperor Theodosius had shuttered the pagan temples in 392, in effect making Christianity the Roman Empire's official religion; as the state collapsed, the Church had increasingly taken on the role of mediator between the Latins and the barbarians. The spread of its doctrine and the conversion of many to the Christian faith seemed an affirmation of Rome's prestige and a guarantee of its survival. At the same time, factions within the Church battled to promulgate new orthodoxies. Such influential thinkers as Augustine bitterly denounced what they saw as religious heresies threatening order and unity. To save the Church and preserve the social order, they mandated strict obedience to established authority and the subordination of civil to spiritual power.

The writings of early Christian leaders, however, had relatively little to say about the natural sciences. Science seemed largely irrelevant to the great theological controversies of the time. In such an environment, scholars—particularly those associated with the Alexandrian schools at the Mediterranean edge of Egypt—dedicated themselves to mathematical studies and the revival of traditions going back centuries, to Plato's Academy in Athens. This intellectual and philosophical pursuit of scientific rationality, reflected in an orderly, predictable celestial realm, contrasted sharply with the social, political, and theological tumult of the times.

Among the writings that survive from that period, one book concerning astronomical matters achieved particularly wide circulation and exerted considerable influence for many centuries thereafter. Written around the year 400 by Ambrosius Theodosius Macrobius, the book owed much of its success to its introduction of the critical analysis of classical texts to a much wider audience than the narrow coterie of specialists normally interested in such matters. Harking back to Plato and the mysticism and numerology of Pythagoras and his followers, Macrobius popularized a strongly mathematical vision of the universe as an orderly, well-regulated celestial system. Though very little is known about this writer, hints in the extant fragments of his work suggest that he was both an accomplished scholar and a high government official.

Macrobius took as his starting point an episode described by Cicero, the Roman statesman, commentator, and famed orator of the first century B.C., in a piece called *Somnium Scipionis* (Scipio's dream), which was dramatically set against the backdrop of the thick river of stars known as the Milky Way. Using passages from this work as opportunities for lengthy excursions into Neoplatonic philosophy, Macrobius ventured into such matters as Pythagorean number lore, world geography, and celestial harmonies. He presented what remained for many centuries the stock picture of the heavens: a spherical Earth located at the center of a spherical universe, encircled by seven planetary spheres and by a celestial sphere that rotates daily from east to west. He also took particular care to explain the Pythagorean doctrine that numbers underlie all physical objects and phenomena.

"Now it is well known," Macrobius wrote, "that in the heavens nothing happens by chance or at random, and that all things above proceed in orderly fashion according to divine law. Therefore it is unquestionably right to assume that harmonious sounds come forth from the rotation of the heavenly spheres, for sound has to come from motion, and Reason, which is present in the divine, is responsible for the sounds being melodious."

A century later, Boethius reinforced this idea in *The Consolation of Philosophy:* "If you wish to discern the laws of the high and mighty

God . . . look up to the roof of highest heaven. There the stars, united by just agreement, keep the ancient peace. . . . Thus mutual love governs their eternal movement and the war of discord is excluded from the bounds of heaven."

Gradually these ideas acquired a more mathematical cast, and scholars passed them on to succeeding generations of students. During the fourth decade of the thirteenth century, no institution of higher learning held a greater appeal for ambitious students than the University of Paris. Here the most prominent teachers of the age carried the fine art of logical argument to its most brilliant pinnacle. Their lectures, commentaries, and debates reverberated throughout the Christian world.

It was perhaps inevitable that the clever and cocky Roger Bacon, barely 20 years old, would journey from Oxford across the English Channel to continue his education and pursue a scholarly career in Paris. Born in 1220 to a wealthy English family, which later lost its fortune by supporting the losing side in a political dispute, Bacon had obtained a thoroughly classical education. Starting at the age of seven or eight, he had studied the Latin authors, including the orator Cicero and the philosopher and statesman Seneca. Five years later, at the University of Oxford, he received training in geometry, arithmetic, music, and astronomy, which constituted the core of an education in the liberal arts.

Bacon was no anomaly. With the gradual emergence of national monarchies, the rapid growth of a wealthy, urban merchant class, and the enduring influence of the Church, universities served both secular and theological purposes. Rulers were interested in fostering the development of an intellectual elite to handle the increasingly complex affairs of burgeoning national states. Ecclesiastical authorities endeavored to reconcile Christian dogma with Greek philosophy and to stamp out the various heresies that threatened its spiritual hold on the populace. The thirteenth century saw not only the building of magnificent, towering cathedrals at Amiens, Rheims, Chartres, Toledo, Cologne, and Salisbury but also the founding of important universities at Cambridge, Padua, Salamanca, Toulouse, and elsewhere. The long-established University of Paris, however, predated these fledgling institutions, and its cutting-edge reputation attracted scholars and students from across Europe.

Arithmetic together with music, astronomy, and geometry formed the medieval quadrivium. In this page from Gregor Reisch's *Margarita philosophica*, published in 1503, Arithmetic is demonstrating to figures representing Pythagoras and Boethius the difference between using counters and Hindu-Arabic numerals for computations. (Library of Congress.)

Despite the monolithic, restrictive world of Church-mandated thought that many associate with this period, medieval Europe tolerated, to various degrees, a remarkably wide range of views. Both secular scholars and theologians passionately argued their viewpoints, often drawing vastly different conclusions from the same source material. Rival religious orders, though still under the umbrella of the same Church, often held opposing beliefs.

For example, the philosophy and physics of Aristotle, first propounded more than 15 centuries earlier, were intricate enough to accommodate many divergent opinions as to their true meaning. There was no single, officially sanctioned form of Aristotelianism. In fact, the Church in 1210 had banned the teaching of Aristotle's philosophy because it apparently contradicted the Church's own teachings. In succeeding decades, however, scholars managed to bring the sacred and the profane into reasonable accord by means of a variety of subterfuges and strategems.

Aristotle's physics, for instance, was both broad and complex. Aristotle imagined the universe as a set of nested spheres centered around Earth, which carried the sun, the moon, and the five known planets. He supposed that different laws governed the celestial and terrestrial realms, and he expressed the view that matter on Earth is composed of four fundamental elements: earth, air, water, and fire. Earth and water had a natural tendency to move toward the center of the universe, whereas air and fire just as naturally moved away. Indeed, Earth had to be a globe because bits of heavy terrestrial material dropped into the center of the universe would naturally pile up in the form of a sphere. The direction and speed of a moving object depended on its precise composition, with heavier bodies falling more rapidly than lighter ones. Denying the existence of atoms, Aristotle also assumed that matter is continuous.

This formulation left much room for debate and invention. Thirteenth-century scholars, for example, expended a great deal of effort debating the tricky question of what happens when a projectile flies through the air. If matter is continuous, how does a body get through? Does the air part in front of it, pulling it forward and closing in behind it, thereby forcing the body ahead?

Some natural philosophers suggested alternative models of the universe, emphasizing different aspects of the world in an effort to reconcile Aristotle's ideas with the results of what were in effect simple experiments and observations of natural phenomena. These ranged from the evident bending of light at the boundary between air and water and between air and glass to the attraction between a magnet and an iron ball and the behavior of a compass needle.

In the early part of the thirteenth century, Robert Grosseteste, one of the University of Oxford's first chancellors, described a compelling,

Variations on Aristotle's theory that the elements of matter consist of earth, air, fire, and water survived for centuries, as illustrated in this diagram from the late fifteenth century. (Library of Congress.)

unified vision of a universe in which light was the fundamental force. Differing with Aristotle and others who had argued that heavenly bodies consisted of some pure, quintessential substance unlike any found on Earth, Grosseteste maintained that the stars themselves are composed of the four earthly elements. Like the attraction of a lump of iron to a magnet, light from the stars attracts comets, which themselves are a form of purified fire. His model also explained the moon's influence on the tides in terms of the way its light causes water to swell and move upward.

Some Aristotelian natural philosophers put great stock in understanding nature through mathematics. Others emphasized the importance of empirical observation. Roger Bacon, in particular, came to symbolize the conviction that both of these approaches were important in comprehending nature, though he was not alone in arguing for the importance of experiment and mathematics—especially number—in helping to unveil the inner workings and deepest secrets of the universe.

It's probable that Bacon received his master of arts degree at the University of Paris, and from 1241 to 1246 he lectured there, like other professors in the faculty of arts expounding on various aspects of Aristotle's works. For an inquisitive mind, it was a time of relative freedom. Moreover, the succession of crusades that brought armies of Europeans to the Near East and other Mediterranean lands resulted in close contact, despite the hostilities, with Islamic customs, culture, and learning. Through their translations of Greek texts and their own original contributions, Arabic authors helped Europeans regain an appreciation of astronomy and mathematics.

Leonardo of Pisa (Fibonacci), for example, learned of arabic numerals firsthand during his travels along the Mediterranean's edge, and in 1202 he wrote an influential book introducing this remarkably economical system of notation to Christian Europe. The same book also included a broad examination of fractions and demonstrated the uses of arithmetic in solving a variety of practical problems, including many linked to commerce. A later treatise presented rules for calculating areas and volumes and for dividing geometrical figures into parts.

During roughly the same period, John of Holywood (Sacrobosco), of the University of Paris, explicitly stated and explored the rules of arithmetic in a slim volume titled *Algorithmus*. In another work, the same scholar provided the first readily accessible explanation by a European of Ptolemy's mathematical model of the solar system, as originally depicted in the *Almagest*. His treatise rejected the widely held notion that the planets were somehow attached to revolving, crystal spheres that had a physical presence. Instead, he favored Ptolemy's simpler, conceptually more elegant, mathematical point of view, which so faithfully reproduced planetary movements.

Bacon undoubtedly encountered many of these provocative ideas during his sojourn in Paris. Restless, energetic, and prone to sharp criticism or dismissal of those who didn't share his views, Bacon returned to Oxford in 1247. There he immersed himself in a stimulat-

SPHAERA IOANNIS

quam nobis. Cuius rei causa est tantum tumor terræ.

Quòd etiam terra habeat tumorem à Septentrione in austrum, & econtrario. sic patet, hominibus existentibus versus Septẽtrionem quædam stellæ sunt sempiternæ apparitionis, illæ. s. quæ ꝓpinquæ accedunt ad polũ arcticum, aliæ aũt sunt sempiternæ occultationis. sicut illæ q̃ sunt propinquæ polo antarctico. Si igitur aliquis ꝓcederet à Septentrione versus austrum, in tantũ posset procedere, ꝙ stellæ q̃ prius erant ei sempiternæ apparitionis, et iam tẽderent in occasum, & quãto magis accederet ad austrũ, tanto plus mouerentur in occasum. Ille iterũ idem hõ posset videre stellas, q̃ prius fuerũt ei sempiternæ occultationis, et ecõuerso cõtingeret alicui procedenti ab austro versus septentrionẽ. Huius aũt rei causa est tantum tumor terræ. Itẽ si terra esset plana ab oriente in occidẽtem, tam cito orirentur stellę occidentalibus, q̃ orientalibus, quod patet esse falsum.

Itẽ si terra esset plana à septẽntrione in austrum, & ecõtrario, stellæ q̃ essent alicui sempiternę apparitiõis, semper apparerẽt ei quocũq; ꝓcedet, quod falsum est, sed ꝙ plana sit, præ nimia eius quantitate, hoĩm visui apparet.

DE SACROBVSTO. 11

QVOD AQVA SIT ROTVNDA.

Quod aũt aqua habeat tumorem, & accedat ad rotũditatem sic patet. Ponatur signũ in littore maris, & exeat nauis à portu, & intantum elongetur, ꝙ oculus existẽs iuxta pedem mali nõ possit videre signũ: stante vero naui, oculus eiusdem existentis in summitate mali bene videbit signum illud, sed oculus existẽtis iuxta pedem mali melius deberet videre signum, q̃ qui est in summitate mali; sicut patet per lineas ductas ab vtroq; ad signum, et nulla alia huius rei causa est, quam tumor aquæ. Excludantur .n. omnia alia impedimẽta sicut nebulæ, et vapores ascẽdentes. Item cum aqua sit corpus homogeneũ, totũ cum partibus eiusdem erit rõnis, sed partes aquę: sicut in guttulis, & roribus herbarum accidit: rotundam naturaliter appetunt formam, ergo & totum cuius sunt partes.

QVOD TERRA SIT CENTRVM MVNDI.

QVOD aũt terra sit in medio firmamenti sita, sic patet. existentibus in superficie terræ stellæ apparent eiusdem quãtitatis, siue sint in medio cæli, siue iuxta ortum, siue iuxta occasum, & hoc ideo quia æqualiter terra distat ab eis. Si .n. terra magis accederet ad firmamẽtum in vna parte q̃ in alia sequeretur ꝙ aliquis existens in illa parte superficiei terræ, quę magis acce-

B iij deret

Around the year 1220, John of Holywood (Sacrobosco) wrote *Sphaera mundi* (Spheres of the world) to introduce ideas from Ptolemy's *Almagest* to medieval Europe. His slim volume explained lunar and solar eclipses and presented evidence that Earth is indeed a sphere. The pages shown come from a 1577 edition. (Library of Congress.)

ing environment quite different from the intense competitiveness and rivalries that typified scholarship in Paris. Strongly influenced by the radical notions of Grosseteste and his disciples at Oxford, Bacon plunged wholeheartedly into various aspects of mathematics, optics, astronomy and astrology, and alchemy.

The study of languages, in particular, became a crucial element in his program. Writing two decades later, Bacon would recall an incident indelibly imprinted in his memory: Once, while lecturing on a classical text at the University of Paris, his Spanish students had laughed at him for mistaking a Spanish word for an Arabic one. Bacon never forgot this lesson, and he zealously pursued accurate translation from sources as close as possible to the original. Demonstrating his dedication to the relentless pursuit of the truth, he wrote: "For, during the twenty years in which I have labored specially in the study of wisdom, after disre-

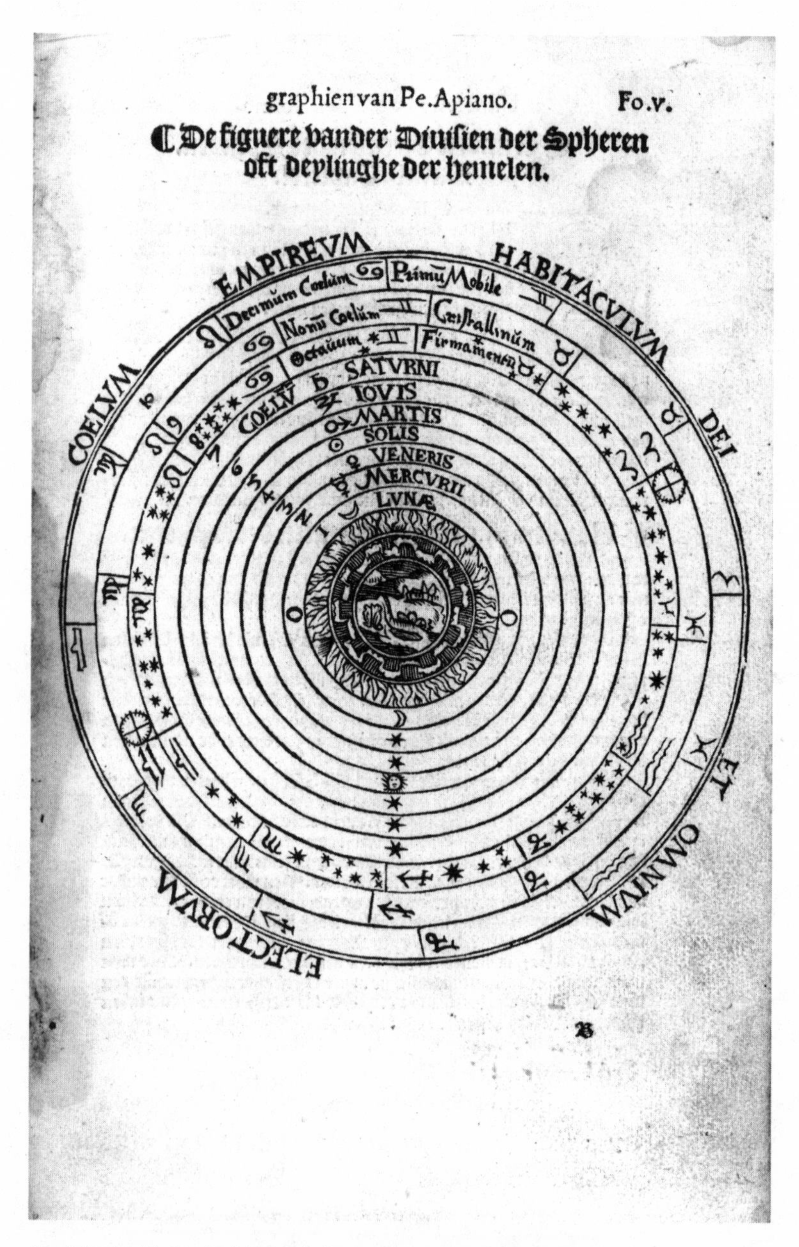

In this 1545 depiction of Aristotelian cosmology, Petrus Apianus shows a stationary Earth at the center of the universe, surrounded by seven concentric "crystalline" spheres, which carry the sun, the moon, and the planets. (Library of Congress.)

garding the common way of thinking, I have put down more than two thousand pounds for secret books and various experiments, and languages and instruments and tables and other things; as well as for searching out the friendships of the wise, and for instructing assistants in languages, in figures, in numbers, and tables and instruments and many other things."

But these were also troubled years for Bacon, in which episodes of feverish activity were interspersed with lengthy visits to Paris, quarrels with various authorities, and ventures into mysticism. In 1257, Bacon became a Franciscan friar but fell afoul of new regulations that forbade members of the order to publish works of any kind without the consent of the proper authorities. Plagued by ill health, he believed himself abandoned and forgotten.

Bacon nonetheless endeavored to overcome the strictures imposed on his writings by appealing to Cardinal Guy de Foulques, an acquaintance from his days in Paris. When the Cardinal was elected Pope Clement IV in 1265, he invited Bacon to submit a copy of his philosophical writings in secret so that their importance could be evaluated. But Bacon had nothing ready, and he spent the next two years frantically compiling the equivalent of an encyclopedia in three volumes. The work encompassed a wide swath of knowledge and reflected its author's unique perspective on the nature of wisdom and the need for educational reform.

Bacon identified four obstacles in the pursuit of truth. He condemned reliance on fallible authority, blind acceptance of long-standing custom, belief in uninformed popular opinion, and displays of false wisdom. Wisdom had to be developed through reason, he argued, and reason by itself could not stand unless confirmed by experience. He described two kinds of experience. One could be obtained through mystical contemplation and the other by means of the senses. Scientific instruments could facilitate the latter and mathematics could make it precise.

In statements that now appear prophetic, Bacon contended that "mathematics is the door and the key to the sciences." Referring to optics, he noted that "in the things of the world, as regards their efficient and generating causes, nothing can be known without the power of geometry." Linking mathematics with astronomy and astrology, he wrote: "For the things of this world cannot be made known without a knowledge of mathematics. For this is an assured fact in regard to celestial things, since two important sciences of mathematics treat of them, namely theoretical astrology and practical astrology. The first . . . gives us definite information as to the number of the heavens and the stars, whose size can be comprehended by means of instruments, and the shapes of all and their magnitudes and distances from the earth, and thicknesses and number, and greatness and smallness. . . . It likewise treats of the size and shape of the habitable earth. . . . All this information is secured by means of instruments suitable for these purposes, and by tables and by canons. . . . For everything works through innate forces shown by lines, angles, and figures."

But these were hardly original notions. Astronomers of antiquity and in Bacon's time were well aware that mathematics in general and calculation in particular played an essential role in their science. Improvements in the observation and measurement of celestial events had long been furthering the development of mathematics, just as problems in mathematical prediction had resulted in new observations and measurements. Bacon himself made only minor contributions to the advancement of mathematics and astronomy. But he was notable in arguing for the usefulness of mathematics, not only in the sciences but also in nearly every other type of human activity, from calendar reform to music and geography.

Curiously, despite his posthumous reputation as a "doctor mirabilis," or doctor of wonders, Bacon appears to have performed relatively few experiments himself, and these were mainly in optics. Moreover, his thinking incorporated a strong element of "natural magic," which he believed could be harnessed for practical use.

The last known date in Bacon's life is 1292, when he wrote a theological work. Condemned and jailed sometime between 1277 and 1279 by Franciscan authorities in Paris for "suspected novelties" in his teachings, Bacon had survived a lengthy imprisonment. Undeterred, in this final work he continued his bitter attacks on the scholarly and theological establishments of his day.

Another part of Bacon's legacy, and that of other natural philosophers as well, was a mystical belief in the power of numbers to lay bare the Creator's grand design. He was among those who hoped to find in numbers the key to deciphering the universe and solving its riddles. These individuals firmly believed that mathematics was the appropri-

The fifteenth century saw a strong revival of Greek studies, abetted by new editions and translations of Greek manuscripts. This frontispiece from *Epitoma in Almagestum*, published in 1496, shows its author, Regiomontanus, and a crowned Ptolemy. (Library of Congress.)

ate means to such an end. In fact, scholars of the fourteenth and fifteenth centuries used quite sophisticated mathematical arguments to criticize and create alternatives to Aristotle's cosmology and concept of motion.

This faith in the power of number and geometry carried considerable weight throughout the Middle Ages and into the Renaissance. During the turbulence of the late sixteenth and early seventeenth centuries, Johannes Kepler searched within the same context for evidence of a divine order. He tried to construct his own harmony of the spheres, this time based on a sun-centered system of planets. The planets themselves were inaccessible, but he could capture their movements in figures scrawled on page after page of paper, and he could depict their relative distances in the geometrical diagrams he fashioned out of nested polyhedrons and spheres.

Kepler's contemporary, Galileo Galilei, made an even more explicit connection between mathematics and nature. In a famous passage in *The Assayer* he wrote, "Philosophy is written in that great book—I mean the universe—that forever stands open before our eyes, but you cannot read it until you have first learned to understand the language and recognize the symbols in which it is written. It is written in the language of mathematics and its symbols are triangles, circles, and other geometrical figures without which one does not understand a word, without which one wanders through a dark labyrinth in vain."

Isaac Newton, in many ways as strongly mystical as Roger Bacon and equally driven to uncover the secrets of nature through numerical and alchemical studies, also believed firmly in the power of mathematics to reveal the workings of the world. His deceptively simple mathematical statements of the laws of motion and the law of gravitation captured the essence of how the universe works. In the *Principia*, Newton expressed the importance of anchoring these concepts in physical reality: "Hitherto I have not been able to discover the causes of these phenomena, and I frame no hypotheses; for whatever is not deduced from the phenomena is to be called an hypothesis; and hypotheses, whether metaphysical or physical, whether of occult qualities or mechanical, have no place in experimental philosophy. In this philosophy, particular propositions are inferred from the phenomena, and afterwards rendered general by induction. Thus it was the impenetrability, the mobility, and the impulsive force of bodies, and the laws of motion and of gravitation, were discovered. And to us it is enough that gravity does really exist, and act according to the laws which we have explained, and abundantly serves to account for all the motions of the celestial bodies, and of our sea."

From such rules, it was easy to envision a completely transparent, deterministic world in which the entire past and future lay within reach. In principle, everything was predictable, and the finest detail was accessible to calculation. One could establish a cause or predict an effect. One could construct yesterday's or tomorrow's world from what one knew of today. One could turn the clock backward or forward with ease.

That was enough for Pierre-Simon Marquis de Laplace to posit a doctrine of determinism and to prove in a limited sense that the solar system is stable. But even in Laplace's time there were hints that standard mathematical procedures, no matter how ingeniously applied, were perhaps inadequate to the task of capturing in every detail the motions of the planets and particularly of the moon.

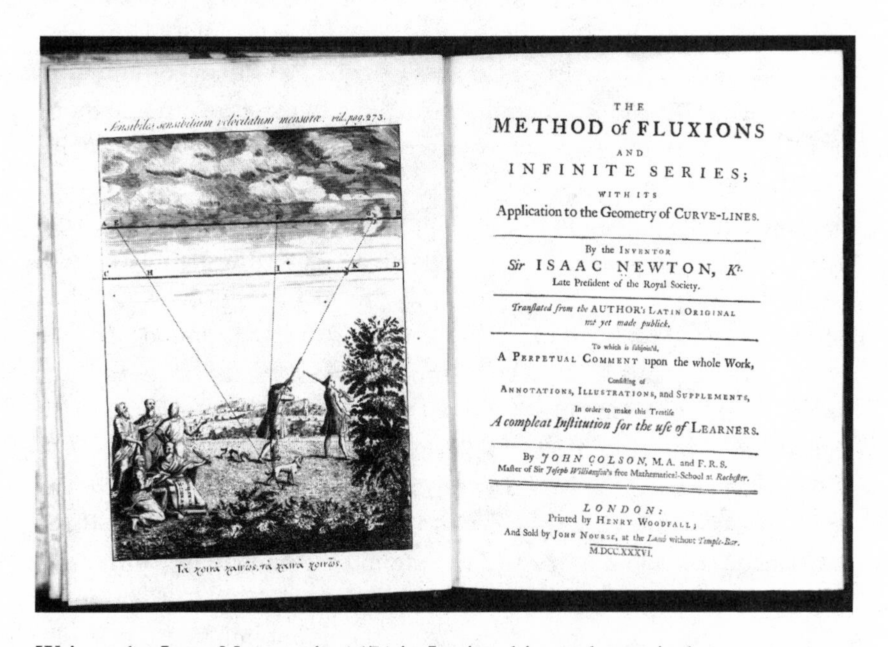

THE
METHOD of FLUXIONS
AND
INFINITE SERIES;
WITH ITS
Application to the Geometry of CURVE-LINES.

By the INVENTOR
Sir ISAAC NEWTON, K^t.
Late President of the Royal Society.

Translated from the AUTHOR's LATIN ORIGINAL
not yet made publick.

To which is subjoin'd,
A PERPETUAL COMMENT upon the whole Work,
Consisting of
ANNOTATIONS, ILLUSTRATIONS, and SUPPLEMENTS,
In order to make this Treatise
A compleat Institution for the use of LEARNERS.

By *JOHN COLSON*, M.A. and F.R.S.
Master of Sir *Joseph Williamson*'s free Mathematical-School at *Rochester*.

LONDON:
Printed by HENRY WOODFALL;
And Sold by JOHN NOURSE, at the *Lamb* without *Temple-Bar*.
M.DCC.XXXVI.

Written by Isaac Newton in 1671 in Latin, this mathematical work, which lays out his ideas for analyzing motion in terms of "fluxions" and infinite series, was not translated and published until 1736, nine years after his death. (Library of Congress.)

Newton's starkly simple equations—his magnificent clockwork—encompassed vast realms of dynamical behavior that would take centuries more of investigation and calculation to bring to light. Henri Poincaré, 200 years after the publication of the *Principia,* led the way by pointing out that deterministic motion, as described mathematically by the differential equations representing Newton's laws, has within it the seeds of uncertainty. Poincaré opened the window on a mathematics of such complexity, with its ample room for the unpredictable, that the harmonies of the spheres again lay hidden, this time within a dense mathematical tangle.

What was astonishing about Poincaré's startling revelation was that the equations themselves remained completely unaltered. Poincaré merely expanded the ways in which they could be interpreted to encompass what at first glance looked like erratic, random movements. This intrinsic unpredictability had been built into the equations from the start. Whether Newton had any inkling that a sensitive dependence on initial conditions lurked in his equations, no one knows. But it was there all along, waiting for the right kind of mind to discover it.

In his 1960 essay, "The Unreasonable Effectiveness of Mathematics in the Natural Sciences," physicist Eugene P. Wigner commented: "The mathematical formulation of the physicist's often crude experience leads in an uncanny number of cases to an amazingly accurate description of a large class of phenomena. This shows that the mathematical language has more to commend it than being the only language which we can speak; it shows that it is, in a very real sense, the correct language." Wigner went on, "Certainly, the example of Newton's law [of gravity], quoted over and over again, must be mentioned first as a monumental example of a law, formulated in terms which appear simple to the mathematician, which has proved accurate beyond all reasonable expectations."

We can also be thankful that the solar system in which we live has been unreasonably kind throughout the long history of human efforts to understand its dynamics and to extend that knowledge to the rest of the universe. At each step along the way, it has served as a perspicacious teacher, posing questions just difficult enough to prompt new observations and calculations that have led to fresh insights, but not so difficult that any further study becomes mired in a morass of confusing detail. As Philip Morrison, emeritus professor of physics at the Massachusetts Institute of Technology, has remarked, "The world is always one step out of our grasp. It's not ten steps out, because then we'd give up. It's not within our grasp, because then we'd be finished. It's just out there at the edge, and that's very desirable."

So it was that the languid movements of a handful of stars distinguished them from the rest. Out of these extraordinary movements of tiny spots of light wandering across the night sky, Aristotle, Hipparchus, and Ptolemy constructed a solar system—and a universe as well—of which Copernicus stumbled upon the true heart. From a tiny discrepancy between calculation and observation, Kepler turned circles into ellipses, and Newton found in Kepler's laws the evidence he needed to support his own formulation of the dynamical laws that apparently govern the universe. As observations improved, new, smaller deviations from the expected appeared, leading to the discovery of additional planets. In the early part of the twentieth century, Albert Einstein's formulation of the general theory of relativity aptly demonstrated that "laws" and their mathematical formulations can change.

In the mathematics used to provide an approximation of the dynamics of the real solar system, Henri Poincaré discovered a remarkably pervasive sensitive dependence on initial conditions—the hallmark of what we now term deterministic chaos. In the context of the solar system, however, it has proved a surprisingly subtle chaos, permitting predictability on a human time scale. Kepler's ellipses suffice for describing a planet's trajectory for several months or a year. With the help of computers and perturbation theory, we can extend this to a few thousand years. Indeed, historians can use tables of planetary and lunar positions to date historical events tied to particular astronomical phenomena. Engineers can plot a spacecraft's orbit and send it on a multiyear mission to the outer planets; with the help of a few judiciously applied course corrections, it will arrive at its target on time. Once a mysterious art, the determination of comet and asteroid orbits is now within the reach of even amateur astronomers, who can accomplish in seconds with desktop computers what it once took Edmond Halley months to do.

Science progresses most rapidly when there is a tight interplay among theory, observation, and experiment. The last decade has seen three developments that promise significant advances over the next

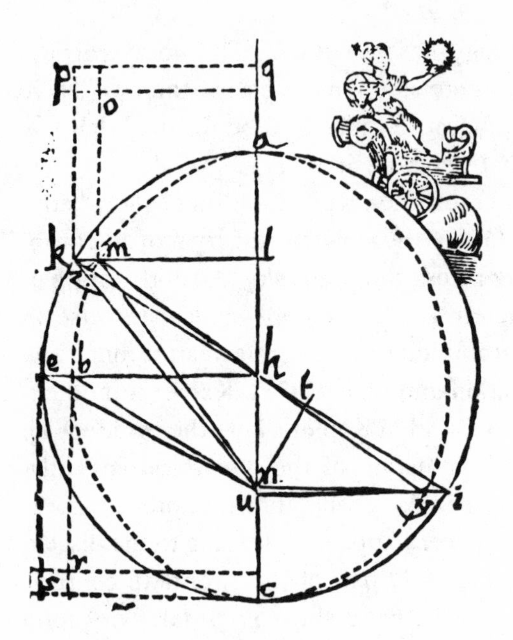

Kepler used this diagram, featuring a "Victorious Astronomy," to demonstrate his two laws of planetary motion. He could thus explain mathematically why the planets changed their velocities as they moved closer to or farther from the sun. (Library of Congress.)

10 years in celestial mechanics and in the understanding of our own corner of the Milky Way. For the first time, scientists have at hand accurate determinations of the masses and distances of all the planets. They have access to a variety of high-speed computers. They can learn from the extensive work that has been done in both mathematics and physics on the behavior of mathematical equations and physical systems that display chaos. With such tools, they are poised for a broadly based theoretical attack on the long-term stability of the solar system. Ultimately, they may achieve a deeper understanding of how the sun and Earth came to be and of whether planets like ours exist elsewhere in the galaxy. In celestial mechanics, this marks a dramatic shift from the dreary though crucial computations of the details of planetary motions and spacecraft trajectories that until recently took up so much of a mathematical astronomer's time.

It should be noted, however, that in mathematical astronomy, especially in studies of the solar system, two distinct traditions operate side by side with surprisingly little interaction. On one side are the planetary scientists, who use the tools of celestial mechanics to help unravel specific problems of the solar system itself. These include the behavior, origin, and evolution of satellites and of the systems of rings around planets. In many cases, these researchers consider not just gravity but also such effects as the solar wind and electromagnetic forces. They tend to publish their results in such scholarly journals as *Icarus*.

The second group focuses much more on solving Newton's equations of motion in idealized situations, where gravity serves as the only force and planets and satellites are considered to be no more than masses concentrated at single points. These researchers often study hypothetical arrangements and numbers of planets just to learn what the mathematics has to say about them. They place a greater emphasis than the planetary scientists do on rigor and mathematical proof. Although they can definitively pinpoint chaotic motion in various special cases, they are far more cautious in attributing it to the solar system at large. Their findings appear in such journals as *Celestial Mechanics*.

This sharp division between the pure and applied traditions in mathematical astronomy inevitably produces friction. One side argues that the other lacks rigor and jumps to conclusions on the flimsy basis of unreliable, ill-considered computer simulations. The other side insists that the purists can't solve problems sufficiently realistic to provide true insights into the solar system's origin, behavior, and evolution. Nonetheless, mathematics and physics are coming closer together again. Philip Holmes, who works at this interface, has remarked: "There is a great ferment of excitement and activity. The artificial distinction between pure and applied mathematics is weakening. Mathematicians and scientists from different fields are talking to one another. Some are even listening."

Increasingly, however, scientists, engineers, and other consumers of mathematics confront situations in which mathematics, even in principle, can't supply definitive answers. Ironically, mathematics itself contains the seeds of this uncertainty. There are fundamental limits to what can be achieved using the tools of mathematics. As recounted in the previous chapters, this realization is just beginning to seep into celestial mechanics. More and more, researchers must confront the implications not just of chaos but of the plethora of other complexities that can reside in the simple mathematical equations used to describe physical systems.

We now have at least one piece of observational evidence on a human time scale that, like the underlying mathematics, suggests that nature, too, can display a sensitive dependence on initial conditions. Saturn's erratically tumbling satellite Hyperion defies any attempt to pinpoint its attitude in space from one month to the next. Asteroid orbits and the sometimes strange configurations of planetary rings also hint at a wonderfully rich and intricate dynamics, but the time scales involved are much longer. And it's useful to keep in mind that there is much more to dynamics than the initial conditions that happen to lead to chaos. Complicated dynamics are not necessarily unpredictable.

With new, more precise data on the masses and positions of the planets and with increasingly powerful computers and algorithms, mathematical astronomers and planetary scientists now have the opportunity to widen the accessible dynamical realm. By mapping the domains of chaos and establishing where it exists in the solar system's phase space, they can begin to explore what it is that sets those limits. Undoubtedly there will be surprises, as there have already been in the movements of Pluto and other planets. And they will also encounter situations in which computation by itself doesn't get them very far, and they will have to resort to describing dynamics in terms of probabilities and qualitative measures.

How well our mathematical models correspond to the physical world in particular cases, however, remains arguable. Does mathematics really have anything meaningful to say about the planet on which we ride so complacently and about the behavior of the solar system?

Minute effects usually ignored in mathematical models of the solar system might prove important in speculations about its long-term future. The solar wind of particles and radiation emanating from the sun carries away mass. Tides on Earth caused by the moon's proximity dissipate energy. Frictional forces between the dense, gaseous atmosphere of Jupiter and its satellites produce a similar effect. Under these influences, planetary and lunar orbits slowly change over millions of years, gradually drifting apart. Such changes may bring them closer to

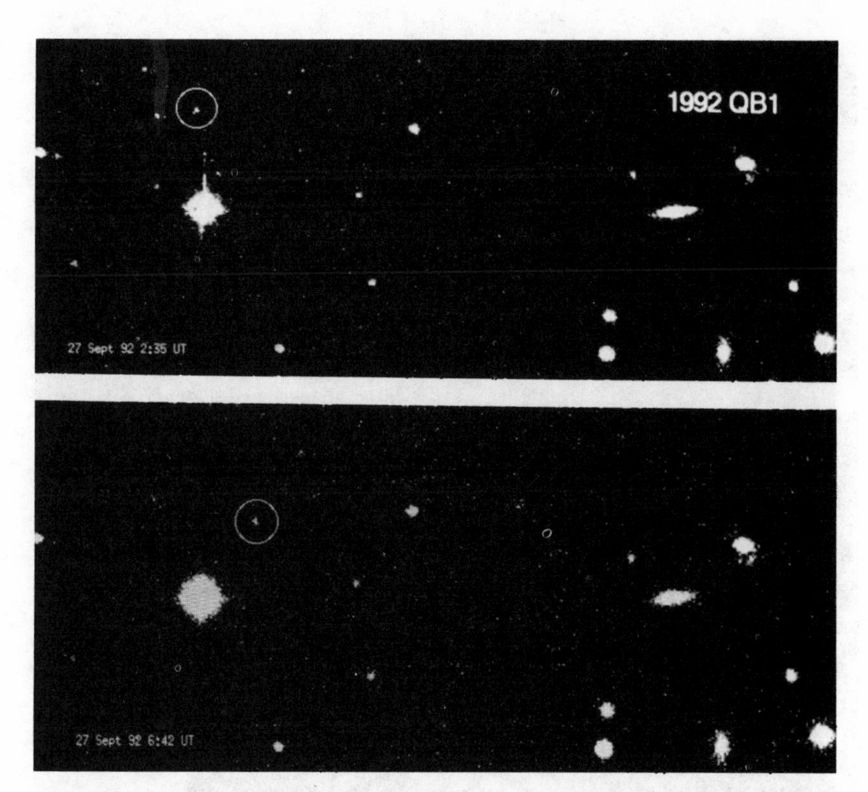

The fringes of the solar system remain largely unexplored. This pair of photographs shows what is probably the most distant object known in the solar system (circled). Discovered in August of 1992 and provisionally designated 1992 QB1, this planetoid has an estimated diameter of 200 kilometers and apparently follows an orbit just outside that of Pluto. (European Southern Observatory.)

dynamical conditions in which chaos can occur and in which drastic alterations become conceivable.

We face a curious situation. As our ability to detect chaos in the motion of real objects improves, it becomes increasingly difficult to assess the relevance of chaos to anything other than the establishment of a time horizon beyond which prediction becomes virtually meaningless. Meanwhile, fundamental questions in celestial mechanics remain unanswered: How much of a role did chaos play in the formation of the solar system? Did the solar system settle down into its present configuration, with well-spaced planets following nearly circular orbits lying in roughly the same plane, within its first few million years? Or has it gradually evolved to its present configuration over the

This series of sketches shows the major stages in the birth of the solar system spanning 100 million years. Terrestrial planets accrete from rocky material in the warm inner regions of the solar nebula. Meanwhile, the huge, gaseous Jovian planets form in the cold outer regions. *Top:* The solar nebula in its initial stages. *Middle:* The early solar system after 50 million years. *Bottom:* Planetary formation is nearly complete after 100 million years. (Reprinted with permission from William J. Kaufmann, *Universe*, 3d ed. New York: W. H. Freeman, 1991, p. 144.)

last five billion years? Were there other planets that have since been ejected? What is Earth's true trajectory? Is it gradually nearing the sun, eventually to be swallowed up, or is it slowly drifting away into the depths of interstellar space?

In the religious language that seems so often to accompany scientific quests, the Holy Grail of celestial mechanics remains the integration of planetary motions over the solar system's entire 4.5-billion-year lifetime. But will that be enough? Can any such integration really cover all the factors that could conceivably affect a long-term history? Or have we already learned basically all we're going to learn about the solar system's dynamical history?

What seems clear now is that the solar system is, on astronomical time scales, no simple, well-regulated mechanical clock. Constantly changing, infinitely complex, it is truly capable of unexpected behavior. As Jack Wisdom has noted, "The solar system has to be recognized for what it is, just another dynamical system, and as such, the discovery that chaotic behavior plays a role in numerous situations in the solar system should come as no surprise."

What we cannot yet fathom is precisely how unexpected that behavior can be. With the solar system's distant past and future effectively hidden from us, we can only speculate about its ultimate dynamical fate—though we suspect that the dying sun's fury will probably render that question moot sometime in the next five billion years or so.

A deep-seated puzzle lies at the heart of this newly discovered uncertainty in our knowledge of the solar system. Was it an accident of celestial mechanics that the solar system happens to be simple enough to have permitted the formulation of Kepler's laws and to ensure predictability on a human time scale? Or could we have evolved and pondered the skies only in a solar system afflicted with a mild case of chaos? Are we special, or were we specially fortunate?

The voyage of discovery into our own solar system has taken us from clockwork precision into chaos and complexity. This still unfinished journey has not been easy, characterized as it is by twists, turns, and surprises that mirror the intricacies of the human mind at work on a profound puzzle. Much remains a mystery. We have found chaos, but what it means and what its relevance is to our place in the universe remains shrouded in a seemingly impenetrable cloak of mathematical uncertainty.

Bibliography

•

General

Beatty, J. Kelly, and Andrew Chaikin. *The New Solar System.* Cambridge, Mass.: Sky Publishing, 1990.

Bell, E. T. *Men of Mathematics.* New York: Simon and Schuster, 1937.

Boyer, Carl B. *A History of Mathematics.* New York: John Wiley, 1968.

Bruno, Leonard C. *Landmarks of Science: From the Collections of the Library of Congress.* New York: Facts on File, 1989.

Campbell, David K. (ed.). *CHAOS/XAOC: Soviet-American Perspectives on Nonlinear Science.* New York: American Institute of Physics, 1990.

Ekeland, Ivar. *Mathematics and the Unexpected.* Chicago: University of Chicago Press, 1988.

Gingerich, Owen. *The Great Copernicus Chase and Other Adventures in Astronomical History.* Cambridge, Mass.: Sky Publishing, 1992.

Gleick, James. *Chaos: Making a New Science.* New York: Viking Penguin, 1987.

Gutzwiller, Martin C. *Chaos in Classical and Quantum Mechanics.* New York: Springer-Verlag, 1990.

Hollingdale, Stuart. *Makers of Mathematics.* London: Penguin Books, 1989.

Kaufmann, William J. *Universe.* 3d ed. New York: W. H. Freeman, 1991.

Kippenhahn, Rudolf (trans. Storm Dunlop). *Bound to the Sun: The Story of Planets, Moons, and Comets.* New York: W. H. Freeman, 1990.

Kline, Morris. *Mathematical Thought from Ancient to Modern Times.* New York: Oxford University Press, 1972.

Kramer, Edna E. *The Nature and Growth of Modern Mathematics.* New York: Hawthorn Books, 1970.

Lang, Kenneth R., and Charles A. Whitney. *Wanderers in Space: Exploration and Discovery in the Solar System.* Cambridge, England: Cambridge University Press, 1991.

Levy, David H. *The Sky: A User's Guide.* Cambridge, England: Cambridge University Press, 1991.

Moser, Jürgen. Is the solar system stable? *The Mathematical Intelligencer* 1, no. 2 (1978):65–71.

Motz, Lloyd, and Jefferson Hane Weaver. *The Story of Physics.* New York: Plenum Press, 1989.

Pannekoek, A. *A History of Astronomy.* New York: Dover Publications, 1989.

Park, David. *The How and the Why: An Essay on the Origins and Development of Physical Theory.* Princeton: Princeton University Press, 1988.

Resnikoff, H. L., and R. O. Wells, Jr. *Mathematics in Civilization.* New York: Dover Publications, 1984.

Rogers, Eric M. *Physics for the Inquiring Mind: The Methods, Nature, and Philosophy of Physical Science.* Princeton: Princeton University Press, 1960.

Roy, A. E. *Orbital Motion.* Bristol, England: Adam Hilger, 1988.

Sambursky, Shmuel (ed.). *Physical Thought from the Presocratics to the Quantum Physicists.* New York: Pica Press, 1974.

Sheehan, William. *Worlds in the Sky: Planetary Discovery from Earliest Times through Voyager and Magellan.* Tucson: The University of Arizona Press, 1992.

Stewart, Ian. *Does God Play Dice? The Mathematics of Chaos.* Cambridge, Mass.: Basil Blackwell, 1989.

Szebehely, Victor G. *Adventures in Celestial Mechanics: A First Course in the Theory of Orbits.* Austin: University of Texas Press, 1989.

Wood, John A. *The Solar System.* Englewood Cliffs, N.J.: Prentice-Hall, Inc., 1979.

1. Chaos in the Clockwork

Olson, Donald W., and Russell L. Doescher. Paul Revere's midnight ride. *Sky & Telescope* 83 (April 1992):437–440.

Olson, Donald W., Russell L. Doescher, and Steven C. Albers. A medieval mutual planetary occultation. *Sky & Telescope* 84 (August 1992):207–209.

Shinbrot, Troy, Celso Grebogi, Jack Wisdom, and James A. Yorke. Chaos in a double pendulum. *American Journal of Physics* 60 (June 1992):491–499.

2. Time Pieces

Brewer, Bryan. *Eclipse.* Seattle: Earth View, 1978.

Brumbaugh, Robert S. *Ancient Greek Gadgets and Machines.* New York: Thomas Y. Crowell, 1966.

Gingerich, Owen (ed.). *The Nature of Scientific Discovery: A Symposium Commemorating the 500th Anniversary of the Birth of Nicolaus Copernicus.* Washington, D.C.: Smithsonian Institution Press, 1975.

Gingerich, Owen. Astronomy in the age of Columbus. *Scientific American* 267 (November 1992):100–105.

Heath, Thomas L. *Greek Astronomy.* New York: Dover Publications, 1991.

Karo, George. Art salvaged from the sea. *Archaeology* 1 (Winter 1948):179–185.

Lindberg, David C. *The Beginnings of Western Science: The European Scientific Tradition in Philosophical, Religious, and Institutional Context, 600* B.C. to A.D. 1450. Chicago: University of Chicago Press, 1992.

Olson, Donald W. Columbus and an eclipse of the moon. *Sky & Telescope* 84 (October 1992):437–440.

O'Neil, W. M. *Time and the Calendars.* Sydney, Aust.: Sydney University Press, 1975.

Price, Derek J. de Solla. An ancient Greek computer. *Scientific American* 200 (June 1959):60–67.

———. *Gears from the Greeks: The Antikythera Mechanism—A Calendar Computer from ca. 80* B.C. New York: Neale Watson Academic Publications, 1974.

3. Wanderers of the Sky

Arnol'd, V. I. (trans. Eric J. F. Primrose). *Huygens and Barrow, Newton and Hooke.* Basel: Birkhäuser Verlag, 1990.

Augarten, Stan. *Bit by Bit: An Illustrated History of Computers.* New York: Ticknor & Fields, 1984.

Bernstein, Jeremy. *Experiencing Science.* New York: E. P. Dutton, 1978.

Donahue, William H. (ed.). *Johannes Kepler's New Astronomy.* Cambridge, England: Cambridge University Press, 1992.

Greco, Vincenzo, Giuseppe Molesini, and Franco Quercioli. Optical tests of Galileo's lenses. *Nature* 358 (9 July 1992):101.

Hanson, Norwood Russell. *Constellations and Conjectures.* Dordrecht, Holland: D. Reidel Publishing, 1973.

Kidwell, Peggy Aldrich. Impermanence enters the astronomical universe: comets, novae and solar radiation, 1400–1650. Presented at the annual meeting of the American Association for the Advancement of Science, Washington, D.C., February 17, 1991.

Koestler, Arthur. *The Sleepwalkers: A History of Man's Changing Vision of the Universe.* London: Hutchinson, 1959.

Lightman, Alan, and Owen Gingerich. When do anomalies begin? *Science* 255 (7 February 1991):690–695.

Ryabov, Y. (trans. G. Yankovsky). *An Elementary Survey of Celestial Mechanics.* New York: Dover Publications, 1961.

Stephenson, Bruce. *Kepler's Physical Astronomy.* New York: Springer-Verlag, 1987.

Thoren, Victor E. *The Lord of Uraniborg: A Biography of Tycho Brahe.* Cambridge, England: Cambridge University Press, 1990.

4. Seas of Thought

Andrade, E. N. da C. *Sir Isaac Newton.* New York: Macmillan, 1954.

Arnol'd, V. I., and V. A. Vasil'ev. Newton's *Principia* read 300 years later. *Notices of the American Mathematical Society* 36 (November 1989):1148–1154.

Bricker, Phillip, and R. I. G. Hughes. *Philosophical Perspectives on Newtonian Science.* Cambridge, Mass.: MIT Press, 1990.

Courant, Richard, and Herbert Robbins. *What is Mathematics?* London: Oxford University Press, 1941.

Dobbs, Betty Jo Teeter. *The Janus Faces of Genius: The Role of Alchemy in Newton's Thought.* Cambridge, England: University of Cambridge Press, 1992.

Erlichson, Herman. Newton's 1679/80 solution of the constant gravity problem. *American Journal of Physics* 59 (August 1991):728–733.

Fauvel, John, Raymond Flood, Michael Shortland, and Robin Wilson (ed.). *Let Newton Be!* Oxford, England: Oxford University Press, 1988.

Hunter, Michael, and Simon Schaffer (ed.). *Robert Hooke: New Studies.* Rochester, N.Y.: Boydell & Brewer, 1990.

Thrower, Norman J. W. *Standing on the Shoulders of Giants: A Longer View of Newton and Halley.* Berkeley and Los Angeles: University of California Press, 1990.

Westfall, Richard S. *Never at Rest: A Biography of Isaac Newton.* Cambridge, England: Cambridge University Press, 1980.

———. The achievement of Isaac Newton: an essay on the occasion of the three hundredth anniversary of the *Principia*. *The Mathematical Intelligencer* 9, no. 4 (1987):45–49.

5. Clockwork Planets

Bell, Eric Temple. *Mathematics: Queen and Servant of Science*. Washington, D.C.: Mathematical Association of America, 1987.

Bollobás, Béla (ed.). *Littlewood's Miscellany*. Cambridge, England: Cambridge University Press, 1986.

Brush, Stephen G. Prediction and theory evaluation: the case of light bending. *Science* 246 (1 December 1989):1124–1129.

Croswell, Ken. The hunt for planet X. *New Scientist* 128 (22–29 December 1990):34–37.

Drake, Stillman, and Charles T. Kowal. Galileo's sighting of Neptune. *Scientific American* 243 (December 1980):74–81.

Henbest, Nigel. Say goodbye to the tenth planet. *New Scientist* 132 (30 November 1991):21.

Hogg, David W., Gerald Quinlan, and Scott Tremaine. Dynamical limits on dark mass in the outer solar system. *The Astronomical Journal* 101 (June 1991):2274–2286.

Lai, H. M., C. C. Lam, and K. Young. Perturbation of Uranus by Neptune: a modern perspective. *American Journal of Physics* 58 (October 1990):946–953.

Gutzwiller, Martin C. The role of perturbation theory in the development of physics. Talk at meeting of the History of Science Society, Seattle, October 26–27, 1990.

Littmann, Mark. *Planets Beyond: Discovering the Outer Solar System*. New York: Wiley, 1990.

Matthews, Robert. Planet X: going, going . . . but not quite gone. *Science* 254 (6 December 1991):1454–1455.

Morgan, Frank. Calculus, planets, and general relativity. *SIAM Review* 34 (June 1992):295–299.

Pierce, David A. The mass of Neptune. *Icarus* 94 (December 1991):413–419.

Temple, Blake, and Craig A. Tracy. From Newton to Einstein. *The American Mathematical Monthly* 99 (June–July 1992):507–521.

Tombaugh, Clyde W. Plates, Pluto, and planets X. *Sky & Telescope* 81 (April 1991):360–361.

Whyte, A. J. *The Planet Pluto*. Toronto: Pergamon Press, 1980.

Will, Clifford M. General relativity at 75: how right was Einstein? *Science* 250 (9 November 1990):770–776.

6. Inconstant Moon

Balfour, Michael. *Stonehenge and its Mysteries.* New York: Scribner's, 1980.

Brecher, Kenneth, and Michael Feirtag (ed.). *Astronomy of the Ancients.* Cambridge, Mass.: MIT Press, 1979.

Gutzwiller, M. C. Chaos and symmetry in the history of mechanics. *Il Nuovo Cimento D* 11 (January–February 1989):1–17.

Hankins, Thomas L. *Jean d'Alembert: Science and the Enlightenment.* New York: Gordon and Breach Science Publishers, 1990.

Hawkins, Gerald S. with John B. White. *Stonehenge Decoded.* Garden City, N.Y.: Doubleday, 1965.

Hide, Raymond, and Jean O. Dickey. Earth's variable rotation. *Science* 253 (9 August 1991):629–637.

7. Prophet of Chaos

Bölling, Reinhard (trans. David E. Rowe). . . . Deine Sonia: a reading from a burned letter. *The Mathematical Intelligencer* 14, no. 3 (1992):24–30.

Briggs, John, and F. David Peat. *Turbulent Mirror: An Illustrated Guide to Chaos Theory and the Science of Wholeness.* New York: Harper & Row, 1989.

Cooke, Roger. *The Mathematics of Sonya Kovalevskaya.* New York: Springer-Verlag, 1984.

Gutzwiller, Martin C. Quantum chaos. *Scientific American* 266 (January 1992):78–84.

Jacobs, Konrad. *Invitation to Mathematics.* Princeton: Princeton University Press, 1992.

Holmes, P. Poincaré, celestial mechanics, dynamical systems and chaos. *Physics Reports* 193 (September 1990):137–163.

Koblitz, Ann Hibner. *A Convergence of Lives: Sofia Kovalevskaia: Scientist, Writer, Revolutionary.* Basel: Birkhäuser, 1983.

Murison, Marc A. The fractal dynamics of satellite capture in the circular restricted three-body problem. *The Astronomical Journal* 98 (December 1989):2346–2359.

Penrose, Roger. *The Emperor's New Mind: Concerning Computers, Minds, and The Laws of Physics.* Oxford, England: Oxford University Press, 1989.

Poincaré, Henri (edited by Daniel L. Goroff). *New Methods of Celestial Mechanics.* New York: American Institute of Physics, 1993.

Schroeder, Manfred. *Fractals, Chaos, Power Laws: Minutes from an Infinite Paradise.* New York: W. H. Freeman, 1991.

8. Band Gaps

Araki, Suguru. Dynamics of planetary rings. *American Scientist* 79 (January–February 1991):44–59.

Bailey, Mark E. Comet orbits and chaos. *Nature* 345 (3 May 1990):21–22.

Beatty, J. Kelly. Galileo calls on Gaspra. *Sky & Telescope* 82 (October 1991):351.

———. A picture-perfect asteroid. *Sky & Telescope* (February 1992):134–135.

Belton, M. J. S., J. Veverka, P. Thomas, P. Helfenstein, D. Simonelli, C. Chapman, M. E. Davies, R. Greeley, R. Greenberg, J. Head, S. Murchie, K. Klaasen, T. V. Johnson, A. McEwen, D. Morrison, G. Neukum, F. Fanale, C. Anger, M. Carr, and C. Pilcher. Galileo encounter with 951 Gaspra: first pictures of an asteroid. *Science* 257 (18 September 1992):1647–1652.

Binzel, Richard P., M. Antonietta Barucci, and Marcello Fulchignoni. The origin of the asteroids. *Scientific American* 265 (October 1991):88–94.

Binzel, Richard P., Shui Xu, Schelte J. Bus, and Edward Bowell. Origins of the near-Earth asteroids. *Science* 257 (7 August 1992):779–782.

Chyba, Christopher F., Paul J. Thomas, and Kevin J. Zahnle. The 1908 Tunguska explosion: atmospheric disruption of a stony asteroid. *Nature* 361 (7 January 1993):40–44.

Cunningham, Clifford J. The captive asteroids. *Astronomy* 20 (June 1992):40–44.

———. Giuseppe Piazzi and the "missing planet." *Sky & Telescope* 84 (September 1992):274–275.

Durda, Dan. All in the family. *Astronomy* 21 (February 1993):36–41.

Esposito, Larry W. Ever decreasing circles. *Nature* 354 (14 November 1991):107.

Gehrels, Tom, and Mildred Shapley Matthews. *Asteroids.* Tucson: The University of Arizona Press, 1979.

Hahn, G., and M. E. Bailey. Rapid dynamical evolution of giant comet Chiron. *Nature* 348 (8 November 1990):132–136.

Holman, Matthew, and Jack Wisdom. Symplectic maps for the *n*-body problem. *The Astronomical Journal* 102 (October 1991):1528–1538.

Kerr, Richard A. Chaotic zone yields meteorites. *Science* 228 (7 June 1985):1186.

———. Galileo's frustrating asteroid pursuit. *Science* 254 (18 October 1991):381–382.

———. Galileo hits its target. *Science* 254 (22 November 1991):1109.

Lecar, Myron, Fred Franklin, and Paul Soper. On the original distribution of the asteroids, IV. Numerical experiments in the outer asteroid belt. *Icarus* 96 (April 1992):234–250.

Lindley, David. A chip off some old rocks. *Nature* 354 (21 November 1991):178.

Marsden, B. G. The computation of orbits in indeterminate and uncertain cases. *The Astronomical Journal* 102 (October 1991):1539–1552.

Matthews, Robert. A rocky watch for earthbound asteroids. *Science* 255 (6 March 1992):1204–1205.

McFadden, Lucy-Ann, and Clark R. Chapman. Interplanetary fugitives. *Astronomy* 20 (August 1992):30–35.

Milani, A., and A. M. Nobili. An example of stable chaos in the solar system. *Nature* 357 (18 June 1992):569–571.

Milani, Andrea, and Zoran Knezevic. Asteroid proper elements and secular resonances. *Icarus* 98 (August 1992):211–232.

Murray, Carl D. Earthward bound from chaotic regions of the asteroid belt. *Nature* 315 (27 June 1985):712.

———. Wandering on a leash. *Nature* 357 (18 June 1992):542–543.

Pollack, James B., and Jeffrey N. Cuzzi. Rings in the solar system. *Scientific American* 245 (November 1981):105–129.

Saha, Prasenjit. Simulating the 3:1 Kirkwood gap. *Icarus* 100 (December 1992):434–439.

Scotti, J. V., D. L. Rabinowitz, and B. G. Marsden. Near miss of the earth by a small asteroid. *Nature* 354 (28 November 1991):287–289.

Steel, Duncan. Our asteroid-pelted planet. *Nature* 354 (28 November 1991):265–267.

Stewart, Ian. Gauss. *Scientific American* 237 (July 1977):123–131.

Talcott, Richard. Galileo views Gaspra. *Astronomy* 20 (February 1992):52–54.

Torbett, Michael V., and Roman Smoluchowski. Chaotic motion in a primordial comet disk beyond Neptune and comet influx to the solar system. *Nature* 345 (3 May 1990):49–51.

Whipple, Arthur L., Paul D. Hemenway, and Doug Ingram. Initial refinement of minor planet orbits for pointing and observation with the Hubble space telescope. *The Astronomical Journal* 102 (August 1991):816–822.

Wisdom, Jack. The origin of the Kirkwood gaps: a mapping for asteroidal motion near the 3/1 commensurability. *The Astronomical Journal* 87 (March 1982):577–593.

———. Chaotic behavior and the origin of the 3/1 Kirkwood gap. *Icarus* 56 (October 1983):51–74.

———. Meteorites may follow a chaotic route to Earth. *Nature* 315 (27 June 1985):731–733.

Yeomans, Donald K. *Comets: A Chronological History of Observation, Science, Myth, and Folklore.* New York: Wiley, 1991.

9. Hyperion Tumbles

Binzel, Richard P., Jacklyn R. Green, and Chet B. Opal. Chaotic rotation of Hyperion? *Nature* 320 (10 April 1986):511.

Coffey, Shannon, André Deprit, Etienne Deprit, and Liam Healy. Painting the phase space portrait of an integrable dynamical system. *Science* 247 (16 February 1990):833–836.

Dermott, Stanley F., Renu Molhotra, and Carl D. Murray. Dynamics of the Uranian and Saturnian satellite systems: a chaotic route to melting Miranda? *Icarus* 76 (December 1988):295–334.

Healy, Liam, and Etienne Deprit. Paint by number: uncovering phase flows of an integrable dynamical system. *Computers in Physics* (September–October 1991):491–496.

Klavetter, James Jay. Rotation of Hyperion, I. Observations. *The Astronomical Journal* 97 (February 1989):570–579.

———. Rotation of Hyperion, II. Dynamics. *The Astronomical Journal* 98 (November 1989):1855–1874.

Marcialis, Robert, and Richard Greenberg. Warming of Miranda during chaotic rotation. *Nature* 328 (16 July 1987):227–229.

Malhotra, Renu. Tidal origin of the Laplace resonance and the resurfacing of Ganymede. *Icarus* 94 (December 1991):399–412.

Murray, Carl D. Chaotic spinning of Hyperion? *Nature* 311 (25 October 1984):705.

Pavelle, Richard, Michael Rothstein, and John Fitch. Computer algebra. *Scientific American* 245 (December 1981):136–152.

Rothery, David A. *Satellites of the Outer Planets: Worlds in Their Own Right.* Oxford, England: Oxford University Press, 1992.

Stewart, John. *Moons of the Solar System: An Illustrated Encyclopedia.* Jefferson, N.C.: McFarland Publishers, 1991.

Tittemore, William C. Chaotic motion of Europa and Ganymede and the Ganymede-Calisto dichotomy. *Science* 250 (12 October 1990):263–267.

Wisdom, Jack, Stanton J. Peale, and François Mignard. The chaotic rotation of Hyperion. *Icarus* 58 (May 1984):137–152.

10. Digital Orrery

Abelson, Harold, Michael Eisenberg, Matthew Halfant, Jacob Katzenelson, Elisha Sacks, Gerald J. Sussman, Jack Wisdom, and Kenneth Yip. Intelligence in scientific computing. *Communications of the ACM* 32 (May 1989):546–562.

Applegate, James H., Michael R. Douglas, Yekta Gursel, Gerald J. Sussman, and Jack Wisdom. The outer solar system for 200 million years. *The Astronomical Journal* 92 (July 1986):176–189.

Freedman, David H. Gravity's revenge. *Discover* 11 (May 1990):54–60.

Hartley, Karen. Solar system chaos. *Astronomy* 18 (May 1990):34–39.

Hut, Piet, and Gerald Jay Sussman. Advanced computing for science. *Scientific American* 257 (October 1987):144–153.

Kerr, Richard A. Pluto's orbital motion looks chaotic. *Science* 240 (20 May 1988):986–987.

Killian, Anita M. Playing dice with the solar system. *Sky & Telescope* 78 (August 1989):136–140.

Murray, Carl. Is the solar system stable? *New Scientist* 124 (25 November 1989):60–63.

Stern, S. Alan, Robert A. Fesen, Edwin S. Barker, Joel W. Parker, and Laurence M. Trafton. A search for distant satellites of Pluto. *Icarus* 94 (November 1991):246–249.

Sussman, Gerald Jay, and Jack Wisdom. Numerical evidence that the motion of Pluto is chaotic. *Science* 241 (22 July 1988):433–437.

Weidenschilling, S. J. A plurality of worlds. *Nature* 352 (18 July 1991):190–191.

Wisdom, Jack, and Matthew Holman. Symplectic maps for the *n*-body problem. *The Astronomical Journal* 102 (October 1991):1528–1538.

11. Celestial Disharmonies

Berger, A., M. F. Loutre, and J. Laskar. Stability of the astronomical frequencies over the earth's history for paleoclimate studies. *Science* 255 (31 January 1992):560–565.

Chernikov, Alexander A., Roald Z. Sagdeev, and George M. Zaslavsky. Chaos: how regular can it be? *Physics Today* 41 (November 1988):27–35.

Ferraz-Mello, S. (ed.). *Chaos, Resonance and Collective Dynamical Phenomena in the Solar System: Proceedings of the 152nd Symposium of the International Astronomical Union Held in Angra dos Reis, Brazil, 15–19 July, 1991.* Norwell, Mass.: Kluwer Academic Publishers, 1992.

Henbest, Nigel. Birth of the planets. *New Scientist* 131 (24 August 1991):30–35.

Kerr, Richard A. Does chaos permeate the solar system? *Science* 244 (14 April 1989):144–145.

———. From Mercury to Pluto, chaos pervades the solar system. *Science* 257 (3 July 1992):33.

Laskar, J. A numerical experiment on the chaotic behaviour of the solar system. *Nature* 338 (16 March 1989):237–238.

———. The chaotic motion of the solar system: a numerical estimate of the size of the chaotic zones. *Icarus* 88 (December 1990):266–291.

Laskar, J., F. Joutel, and P. Robutel. Stabilization of the Earth's obliquity by the Moon. *Nature* 361 (18 February 1993):615–617.

Laskar, J., and P. Robutel. The chaotic obliquity of the planets. *Nature* 361 (18 February 1993):608–612.

Laskar, Jacques, Thomas Quinn, and Scott Tremaine. Confirmation of resonant structure in the solar system. *Icarus* 95 (January 1992):148–152.

Lecar, Myron, Fred Franklin, and Marc Murison. On predicting long-term orbital instability: a relation between the Lyapunov time and sudden orbital transitions. *The Astronomical Journal* 104 (September 1992):1230–1236.

Milani, Andrea. Emerging stability and chaos. *Nature* 338 (16 March 1989):207–208.

Murray, Carl D. Seasoned travellers. *Nature* 361 (18 February 1993):586–587.

Quinlan, Gerald D., and Scott Tremaine. Symmetric multistep methods for the numerical integration of planetary orbits. *The Astronomical Journal* 100 (November 1990):1694–1700.

Quinn, Thomas R., Scott Tremaine, and Martin Duncan. A three million year integration of the earth's orbit. *The Astronomical Journal* 101 (June 1991):2287–2305.

Stern, Alan. Where has Pluto's family gone? *Astronomy* 20 (September 1992):40–47.

Sussman, Gerald Jay, and Jack Wisdom. Chaotic evolution of the solar system. *Science* 257 (3 July 1992):56–62.

Touma, Jihad, and Jack Wisdom. The chaotic obliquity of Mars. *Science* 259 (26 February 1993):1294–1297.

Wisdom, Jack. Urey prize lecture: chaotic dynamics in the solar system. *Icarus* 72 (November 1987):241–275.

12. Machinery of Wonder

Baumgartner, Frederic J. Starry messengers. *The Sciences* 32 (January–February 1992):38–43.

Binzel, Richard P. 1991 Urey prize lecture: physical evolution in the solar system—present observations as a key to the past. *Icarus* 100 (December 1992):274–287.

Boethius (trans. Richard Green). *The Consolation of Philosophy.* Indianapolis: Bobbs-Merrill, 1962.

Kline, Morris. *Mathematics and the Search for Knowledge.* New York: Oxford University Press, 1985.

Ruelle, David. *Chance and Chaos.* Princeton: Princeton University Press, 1991.

Segre, Michael. *In the Wake of Galileo.* New Brunswick, N.J.: Rutgers University Press, 1991.

Stewart, Ian. The symplectic revolution. *The Sciences* 30 (May–June 1990):29–36.

Kleiner, Israel. Rigor and proof in mathematics: a historical perspective. *Mathematics Magazine* 64 (December 1991):291–314.

Wigner, Eugene P. The unreasonable effectiveness of mathematics in the natural sciences. *Communications in Pure and Applied Mathematics* 13 (February 1960):1–14.

Software

Orbits 1.02 (Physics Academic Software). New York: American Institute of Physics.

James Gleick's Chaos: The Software 1.01. Sausalito, Calif.: Autodesk, Inc.

Chaos Demonstrations 1.1 (Physics Academic Software). New York: American Institute of Physics.

Dance of the Planets: Space Travel for the Inquiring Mind 2.5. Loveland, Colo.: A.R.C. Science Simulation Software.

Index